# National Styles
# of Regulation

# Cornell Studies in Political Economy

EDITED BY
PETER J. KATZENSTEIN

# National Styles of Regulation

## Environmental Policy in Great Britain and the United States

### David Vogel

### Cornell University Press

*Ithaca and London*

ACA 6446 - 0/3

*To Virginia,*
*with admiration and love*

# Contents

# *Preface*

This book has three primary objectives: to provide an overview of British environmental policy, to compare the patterns of government regulation in Great Britain and the United States, and to explain why these two nations have adopted such divergent approaches to controlling the externalities associated with industrial growth. Its broader aim is to link the study of government regulation of business with that of comparative politics.

With the exception of the activities enjoined by antitrust legislation, relatively few aspects of corporate behavior that are regulated in the United States are not regulated in other countries also. But the United States remains distinctive in that its rules and regulations tend to be consistently stricter than those of other capitalist countries, and it provides more opportunity for political participation by nonindustry constituencies. Many scholars have attempted to explain why the United States is the only capitalist nation without a socialist movement, but virtually no effort has been made to account for its equally distinctive approach to the regulation of corporate social conduct.[1] This book represents, in part, an effort to fill this gap.

It also attempts to integrate two distinctive bodies of literature that have tended to ignore one another. With only a handful of exceptions, students of government regulation of business in the United States, regardless of their discipline, have shown little interest in comparative public policy. While the issues addressed by regulatory authorities in the United States tend to be similar to those confronted by policy makers in other advanced democracies, until recently only in the area of drug regulation have American scholars displayed much interest in comparing foreign regulatory policies and practices with those of the United States.[2]

9

As a result, students of domestic public policy have been deprived of a unique vantage point from which to understand the dynamics of government regulation of business in the United States. (The obverse is not true: European students of government regulation seem to be relatively familiar with American regulatory policies and procedures. British scholars in particular have written extensively on the contrast between British and American regulatory policies.)[3]

At the same time, students of comparative politics have tended to overlook the politics of regulation of corporate social conduct. We have numerous comparisons of the advanced industrial nations in the areas of industrial policy, social welfare policy, trade policy, monetary and fiscal policy, and the like, but relatively few that examine policies on environmental protection, consumer protection, occupational health and safety, and equal employment opportunity.[4] There is a literature on Japanese environmental policy and growing interest in German environmental policy, but relatively little of this work is comparative in focus.[5]

Before 1980, only one book-length study of environmental regulation across national boundaries had been published by a political scientist: Cynthia Enloe's *Politics of Pollution in Comparative Perspective*.[6] In 1980 Lennart Lundquist published *The Hare and the Tortoise: Clean Air Policies in the United States and Sweden*, and the following year Steven Kelman published *Regulating America, Regulating Sweden*, a cross-national examination of occupational health and safety policies.[7] Since 1980 several articles and reports have summarized the results of an exhaustive study of the formulation and implementation of controls over sulfur dioxide emissions in ten European countries, sponsored by the Institute for Environment and Society in Berlin.[8] In 1982 the results of three major research projects on comparative health and safety regulation were made public: *Environmental Regulation of the Automobile*, published by the Center for Policy Alternatives at MIT; *Chemical Regulation and Cancer*, published by the Program on Science, Technology, and Society of Cornell University; and *Risk Analysis and Decision Processes*, originally published by the International Institute for Applied Systems Analysis in Austria. (The latter two studies have since been released as books.)[9] Harvard Business School has recently published a comparative study of the regulation of vinyl chloride by Joseph Badaracco.[10] And over recent years several more specialized monographs have been issued.[11] This book draws on this small but growing body of literature and seeks to make a contribution to it.

Strictly speaking, this book is not a full-scale comparison of British and American environmental policy. In this sense it differs from the two most important comparative studies of government social regulation by political scientists, those by Lennart Lundquist and Steven Kelman. The first half of the book deals almost exclusively with British environmental policy. A wealth of material has been published on American environmental policy, and little is to be gained by duplicating it here. Since relatively little is known about British environmental policy outside of Great Britain, however, the first three chapters are devoted to a detailed discussion of British environmental regulation, with particular emphasis on the period since the mid-1960s. Only the second half of the book is explicitly comparative. In a sense, this book can be more faithfully described as an examination of British environmental policy as seen through the eyes of a student of American politics. Moreover, while the main focus of the book is on environmental policy, its actual scope is considerably broader. The second half of the book attempts to use the comparative analysis of British and American environmental regulation as a vehicle for generalizing about each nation's distinctive regulatory style and the dynamics of the relationship among business, government, and the public.

My analysis of environmental regulation in both countries is, by necessity, somewhat selective. I deal only peripherally with the controversy over the building of nuclear power plants in either country, since the politics of that issue appears so distinctive and complex as to require a book-length analysis in its own right. The material on Great Britain deals primarily with environmental policy in two areas, pollution control and land-use planning, although it also discusses the control of toxic substances and the controversy over the weight of lorries. Thus other areas of British environmental regulation, such as the control of noise and wildlife protection, receive less emphasis than they may deserve. The American material concentrates on three areas: the control of pollution, the regulation of hazardous substances, and industrial siting.

The Introduction offers a brief survey of the contrasting regulatory styles of the United States and Great Britain. Chapter 1 traces the history of British environmental policy from the 1820s, placing particular emphasis on the upsurge in public concern with this issue in the late 1960s. It also discusses the nature of the British environmental movement and examines its influence on public policy. The next two chapters focus in more detail on two broad dimensions of British environmental policy: Chapter 2 deals with the control of air and water pollution; Chapter 3 discusses the British system of land-use planning. Chapters 4 and 5 are explicitly comparative in focus. Chapter 4 analyzes the effectiveness of British and American environmental regulation and seeks to explain why this issue has produced so much more conflict between business and government in the United States than in Great Britain. Chapter 5 begins to put British and American regulation in a broader context; it compares the patterns of government regulation in the two societies across a wide range of other policy areas. Chapter 6 traces the historical evolution of government–business relations in both societies and offers an explanation as to why their approaches to government regulation of industry have come to diverge so significantly. The last chapter returns to the theoretical issues raised in the Introduction: it seeks to integrate the comparative study of regulation with the literature on both British and American politics and comparative political economy.

This book has benefited significantly from the critical readings given to various versions of the manuscript by persons on both sides of the Atlantic. Their efforts have significantly reduced the number of factual errors in my description of British environmental policy and are largely responsible for whatever analytical coherence the book displays. I trust they will judge the final product worthy of the substantial amount of time and energy they have invested in it. In Britain, all or portions of the manuscript were read by Timothy O'Riordan, Jeremy Richardson, Christopher Wood, Wyn Grant, Lord Ashby, and Anthony Baker. In the United States I benefited from the suggestions of Michael Unseem, Boston University; Jeanne Logsdon, University of Santa Clara; Ian Maitland and Alfred Marcus, University of Minnesota; Peter Hall and Thomas McCraw, Harvard University; and several colleagues at the University of California at Berkeley: Robert Kagen, Eugene Bardach, Richard Abrams, Aaron Wildavsky, John Zysman, and Dow Votaw. In addition, Robert Tricker, Nuffield College, Oxford; Graham Wilson, University of Wisconsin; Michael Asimov, University of California, Los Angeles; James Douglas, Northwestern University; Michael Reagan, University of California, Riverside; and Julian Franks, London Business School, made helpful criticisms of my two previous writings on this subject. I owe a particular debt of gratitude to the editor of the series in which this book appears, Peter Katzenstein. He originally encouraged me to expand my 1982 APSA convention paper into a book and his insightful criticisms have contributed immeasurably to the quality of the final product.

I must also acknowledge another kind of debt—to the British scholars whose work I have relied upon so extensively in preparing a major portion of this study. The embarrassingly rich literature on British environmental regulation produced by British geographers, natural scientists, city and regional planners, and political scientists deserves to be more widely known in the United States. While I cite many of these works at some point in this book, I want to acknowledge here the research of the following scholars, whose efforts have so greatly facilitated my own: Jeremy Richardson, University of Strathclyde; Roy Gregory, University of Reading; Keith Hawkins, Centre for Socio-Legal Studies, Wolfson College, Oxford; Eric Ashby, Clare College, Cambridge; Nigel Haigh, Institute for European Environmental Policy; Wyn Grant, University of Warwick; Anthony Barker, University of Essex; Michael Hill, School for Advanced Urban Studies, University of Bristol; Jane Goyder, Bartlett School of Architecture and Planning, University College, London; Francis Sandbach, University of Kent; Andrew Blowers, Open University; Timothy O'Riordan, School of Environmental Sciences, University of East Anglia; Gerald Rhodes, Royal Institute of Public Administration; Richard Kimber, University of Keele; and Christopher Wood, University of Manchester.

This book could not have been written without the financial assistance of the Center for Law and Economic Studies of Columbia University. I thus owe a large debt to Betty Bock, Harvey Goldschmidt, and Ira Millstein, the directors of the Center's Project on the Large Corporation. I also benefited from the hospitality of the London Business School, where I spent six months as a sabbatical visitor, and of its faculty dean, David Chambers. A grant from the Institute of International Studies of the University of California defrayed the costs of manuscript preparation. I express my thanks to its director, Carl Rosberg, and his excellent staff. Research support was also provided by the Program in Business and Social Policy of the Center for Research in Management of the University of California, Berkeley. Richard Kohl, Department of Economics; David McKendrick, Berkeley Business School; and Susan Erickson and Anthony Daley, Department of Political Science, provided invaluable research assistance.

Two chapters of this book were presented as papers at the annual meetings of the American Political Science Association in 1982 and 1984, and another was presented to the conference "Regulation in Britain," held at Trinity College, Oxford, in September 1983. Earlier versions of parts of the book have appeared as "Cooperative Regulation: Environmental Protection in Great Britain," *The Public Interest*, no. 72 (Summer 1983), pp. 88–106, copyright © 1983 by National Affairs, Inc.; and "Comparing Policy Styles: Environmental Protection in the United States and Great Britain," *Public Administration Bulletin* 42 (1983): 65–78. They are reproduced here in slightly altered form by permission of those journals. In addition, I have presented the arguments developed here at seminars at universities in both the United States and Britain, including the Henley School of Management, the University of Leeds, the University of Essex, Harvard University, the University of Pennsylvania, Cornell University, the University of Minnesota, and the University of California at Riverside, Santa Barbara, and Berkeley. I thank the numerous individuals who attended these presentations for their many valuable criticisms and suggestions.

A year after I returned to the United States from Great Britain, Meinhof Dierkes of the Science Center in Berlin invited me to prepare a paper on the state of the art of cross-national research on environmental policy. This paper was subsequently presented at the conference "Cross-National Policy Research," held in Berlin in December 1983. Both the preparation of this paper and my participation at the Berlin conference provided me with a rare opportunity to understand the links between my own research and that of the other scholars working in this area.

The task of typing and retyping this manuscript—made all the more daunting by my illegible handwriting—fell to four people: Nadine Zelinski, Christine Lundholm, Gwen Cheesberg, and Marcie McGaugh. I can never adequately thank them for their perseverance and good cheer.

I gratefully acknowledge the influence of my parents, Harry and Charlotte Vogel, who laid the foundations for my interest in Europe while I was growing up in New York City.

My greatest thanks go to my family, Virginia, Philip, and Barbara, who accompanied me to Britain and survived the coldest London winter in nearly a century. They now know more about British environmental regulation than anyone in their right mind would ever want to know.

As always, my greatest debt is to my wife, Virginia, whose faith in me during the three years I worked on this book helped sustain my own.

DAVID VOGEL

*Berkeley, California*

# Abbreviations

| | |
|---|---|
| ACP | Advisory Commission on Pesticides |
| BACA | British Agricultural Chemical Association |
| bpm | best practicable means |
| CBI | Confederation of British Industry |
| CEGB | Central Electricity Generating Board |
| CLEAR | Campaign for Lead-free Air |
| CoEnCo | Council for Environmental Conservation |
| CPRE | Council for the Protection of Rural England |
| CSD | Committee on Safety of Drugs |
| DOE | Department of the Environment |
| EDB | ethylene dibromide |
| EEC | European Economic Community |
| EIA | environmental impact assessment |
| EPA | Environmental Protection Agency |
| ETS | emergency temporary standard |
| FDA | Food and Drug Administration |
| FOE | Friends of the Earth |
| FRG | Federal Republic of Germany |
| GNP | gross national product |
| HSC | Health and Safety Commission |
| HSE | Health and Safety Executive |
| HSWA | Health and Safety at Work Act |
| ICI | Imperial Chemical Industries |
| LBC | London Brick Company |
| LEG | liquefied energy gas |
| LNG | liquefied natural gas |
| MP | member of Parliament |
| NIOSHA | National Institute of Occupational Safety and Health |
| NUR | National Union of Railwaymen |

OECD      Organization for Economic Cooperation and Development
OSHA      Occupational Safety and Health Act/Administration
OPEC      Organization of Petroleum Exporting Countries
PCB       polychlorinated biphenyl
ppm       parts per million
PROBE     Public Review of Brickmaking and the Environment
PSD       prevention of significant deterioration
PSPS      Pesticides Safety Precautions Scheme
RTZ       Rio Tinto–Zinc
RWA       regional water authority
SEC       Securities and Exchange Commission
SSSI      sites of special scientific interest
TCDD      2,3,7,8-tetrachlorodibenzo-dioxin
TRG       Transport Reform Group
TUC       Trades Union Council
UK        United Kingdom
VC        vinyl chloride
WARA      Wing Airport Resistance Association
WQO       water-quality objective

*National Styles
of Regulation*

# Introduction

Both the Americans and the British have long histories of concern for the quality of their physical environments: the "countryside" occupies a status in English culture similar to that enjoyed by the "wilderness" in the United States. The Alkali Inspectorate, the central government body responsible for controlling the more complex sources of air pollution, was established by Parliament in 1861, while the framework for American conservation policy dates from the Progressive Era. The efforts of reformers at the turn of the century to enact smoke-control ordinances in the United States paralleled those of the Coal Smoke Abatement Society, established in London about the same time; in fact, civic groups in Pittsburgh described the periodic air inversions that blacked out their city as "Londoners." More recently, the Clean Air Act enacted by Congress in 1963 literally took its name from the statute Parliament had previously approved following the London "killer fog" of 1952.

Over the last twenty years, awareness of environmental issues has increased substantially in both the United States and Great Britain. The wreck of the oil tanker *Torrey Canyon* off the coast of Cornwall in 1966 was followed a year later by the blowout of an oil well off the coast of Santa Barbara. The publication of *Silent Spring* in the United States was followed within two years by a report from the Nature Conservancy in Britain documenting the chemical contamination of the eggs of sea birds. And since the early 1970s the press of both nations has regularly carried reports describing the threat posed by toxic wastes to human health. Many of the issues that have surfaced in Britain—the siting of a third international airport to serve the London area, the mining of natural resources in national parks, the location of energy-related infrastructure, the safety of nuclear energy, the lead content of petrol (gasoline), the

construction of motorways (highways) in rural and urban areas, the
disposal of toxic wastes, the weight of lorries (trucks) permitted on British
ads, the "export" of acid rain—are similar to those that have appeared
he political agenda of the United States.

e the mid-1960s the membership of environmental organizations
r involvement in political activity have increased substantially in
tries.[1] The membership of the Sierra Club increased from
68 to 136,000 in 1972, while that of the National Audubon
arly tripled during the same period. During the same four
he Society for the Promotion of Nature Conservation increased
membership from 35,000 to 75,000 and the membership of the Royal
Society for the Protection of Birds expanded from 41,000 to 108,000.
In addition, new environmental organizations were established in both
societies, including the Conservation Society and the Council on Environmental Conservation in Britain and the Environmental Defense Fund
and Environmental Action in the United States. Two organizations,
Greenpeace and Friends of the Earth, have chapters in both societies.
Environmental organizations in both countries have drawn their support
from roughly the same socioeconomic base—the more affluent and highly
educated—and both movements have worked closely with the media to
make the public more aware of the environmental dimensions of both
corporate and government decisions.

During the first half of the 1970s, each government responded to the
public's increased concern about environmental quality with important
administrative and legislative initiatives. A permanent Royal Commission
on Environmental Pollution was established within a few months of the
Council on Environmental Quality, and the organization of a Department of the Environment in Britain closely followed the establishment
of the Environmental Protection Agency in the United States. Between
1969 and 1974—a period that coincided with the high point of environmental concern in both countries—each nation enacted important
environmental statutes: the National Environmental Policy Act (1969),
the Clean Air Act Amendments (1970), and the Federal Water Pollution
Control Act (1972) were adopted by the United States, and the Deposit
of Poisonous Waste Act (1972), the Water Act (1974) and the Control
of Pollution Act (1974) were enacted by Parliament. All were approved
with substantial bipartisan support, with the most important initiatives
taking place while the more conservative political party was in power in
each country.

Despite these similarities, it is the differences in the strategies of the
two nations for improving the quality of their physical environment and

safeguarding the health of their population that are most striking. Notwithstanding the common roots of their political and legal systems, their approaches to environmental regulation differ from each other more than do those of any other two industrialized democracies.

On balance, the American approach to environmental regulation is the most rigid and rule-oriented to be found in any industrial society; the British, the most flexible and informal. The United States makes more extensive use of uniform standards for emissions and environmental quality than does any other nation; the British, with a handful of exceptions, employ neither. The United States requires the preparation of elaborate environmental impact statements; their use remains optional in Britain. The United States makes virtually no use of industry self-regulation to improve environmental quality; the British rely on it extensively. Regulatory authorities in America take companies to court more frequently than those of any other country; prosecution in Great Britain is extremely rare. The thrust of American environmental regulation has been to restrict administrative discretion as much as possible; in Britain regulatory officials remain relatively insulated from both parliamentary and judicial scrutiny. While environmental regulation in Great Britain has exhibited remarkable continuity over the last three decades, only in Japan has the direction of environmental policy changed as rapidly over the last twenty years as it has in the United States. And while the saliency of environmental issues in different countries has varied over time, over the last fifteen years in no nation has environmental policy been the focus of so much political conflict as it has in the United States.

The most striking difference between the environmental policies of Great Britain and the United States has to do with the relationship between business and government. While in every industrial nation businesses have had to confront an increase in environmental regulation since the late 1960s, no other business community is so dissatisfied with its nation's system of environmental controls as the American business community. In Great Britain, by contrast, the relations between the two sectors have been relatively cooperative. Not only do regulatory officials tend to believe that virtually all companies are making a good-faith effort to comply with environmental regulations, but in the scores of interviews that I conducted with corporate executives in Great Britain, including several with the subsidiaries of American-based multinationals, not one could cite an occasion when his firm had been required to do anything it regarded as unreasonable. In fact, the British business community has been among the most consistent defenders of its nation's system of environmental controls in the face of efforts by the European Economic Community to "harmonize" Britain's regulatory policies with those of other European nations.

In America, environmental regulation has seriously exacerbated tension between business and government: each tends to accuse the other of acting in bad faith. Many American executives and students of regulation blame environmental regulation for many of the difficulties that have confronted the American economy in recent years. It has been accused of reducing productivity, increasing inflation and unemployment, impairing the rate of new capital formation, needlessly delaying important new investments—particularly in the area of energy—creating additional paperwork, and diverting corporate research and development expenditures from productive to nonproductive uses.[2] But while there is no shortage of explanations for the poor performance of British industry in the postwar period, environmental regulation is not among them. Significantly, while both the Thatcher government and the Reagan administration have sought to reduce the burdens of government on industry, only the latter has attempted to make any substantial changes in environmental policy.

Is the relative lack of tension between government officials and industrial managers in Britain due to the latter's "capture" of the former? How do the two countries compare in regard to the effectiveness of environmental regulation? While American rules and regulations enacted since 1969 do demand more from industry, environmental quality has not improved more rapidly in the United States. Instead policy implementation in the United States has become more contentious. While it is difficult to make cross-national comparisons of policy effectiveness, on balance the two nations appear to have made comparable progress in controlling industrial emissions, safeguarding public health, and balancing conservation values with industrial growth.

Between 1958 and 1978, urban ground concentration levels of sulfur dioxide fell by approximately 50 percent in Great Britain, while between 1966 and 1979 they declined by approximately 75 percent in the United States. Between 1958 and 1981 the emissions of smoke from industrial sources declined by 94 percent in Great Britain, while between 1970 and 1977 particulate emissions in America were nearly halved. The restrictions on the burning of coal established by the Clean Air Act (1956) have played a critical role in improving air quality in many of Britain's cities. London's famous fog, which in addition to reducing visibility presented a substantial threat to public health, has now all but disappeared. As a result of the Clean Air Act Amendments of 1967, 1970, and 1977, American automobile emissions have been reduced by 67 percent and air quality has measurably improved in most urban areas.

Britain's water quality has also substantially improved over the last two decades: the Thames, whose foul odors at one time made it difficult for Parliament to conduct its business, is now the cleanest tidal river in the world and contains more than a hundred species of fish. The transfer of responsibility for the control of water pollution from the American states to the federal government has led to a substantial improvement in many of the nation's bays and lakes, including Lake Erie, which, like the Thames, was once considered "dead." Both nations have substantially expanded the amount of land designated as "conservation areas," and each has been able to bring on stream major new sources of energy— Britain in the North Sea off Scotland, America in the North Slope of Alaska—without adversely affecting the ecology of either region.

At the same time, the records of both nations exhibit a number of shortcomings. The United States has experienced more difficulty than Britain in controlling the disposal of toxic wastes and reducing ground-water contamination, while the British record with respect to the safety of its nuclear energy facilities is considerably poorer. On the other hand, Canada's criticisms of the inadequacy of American controls over sulfur emissions from midwestern utilities are similar to those leveled by the Swedes against the British policy of controlling sulfur emissions through the construction of "tall stacks." In addition, both nations contain water-ways that remain highly polluted, and air quality remains poor in parts of the north of England and in the Ohio River Valley.

My argument is *not* that either nation's environmental controls have been effective, but that Britain's emphasis on voluntary compliance has not proved any more—or less—effective in achieving its objectives than the more adversarial and legislative approach adopted by policy makers in the United States. American regulatory policy has been more ambitious, but as a result it has produced greater resistance from business. British regulatory authorities demand less, but because their demands are perceived as reasonable, industry is more likely to comply with them.

Is environmental regulation less contentious in Britain because compliance has been less costly? Over the last decade, both nations have devoted approximately the same share of their gross national product to pollution control. While particular American industries have incurred heavier compliance costs than their British counterparts, they have also been more able to afford them. On balance, business opposition to environmental regulation in the United States has less to do with economics than with politics. It is not that the American system of regulation is an adversarial one because the costs of compliance are so high; rather it is the adversarial nature of American environmental regulation that makes both the direct and indirect costs of compliance appear excessive. It is the way in which environmental policy is made and implemented—not the direct cost of complying with it—that accounts for the resentment it has aroused within the American business community and the relative lack of such resentment on the part of the British business community.

Administrators in Britain enjoy substantially more discretion than their counterparts in the United States. Less bound by fixed standards, they are able to tailor regulations to the particular circumstances of individual firms and industries. Moreover, the rules they issue tend to be based on a consensus among engineers and scientists in both sectors. This is particularly true in the highly controversial area of risk assessment. In America, on the other hand, regulatory officials have often found themselves pressured by the courts and the Congress to make and enforce rules that are perceived as both unreasonable and arbitrary by the firms that have to comply with them. American environmental regulations have tended to be technology-forcing, while British regulatory requirements have been tied to both the technological and financial capacity of industry to comply with them. The British system also imposes fewer administrative and legal costs on industry. Because the pace of policy innovation is more gradual, moreover, the British system creates less uncertainty on the part of corporate planners. In sum, environmental regulation may have similar environmental and economic impacts in the two countries, but its political consequences differ substantially.

The differences between British and American regulatory policies are not confined to environmental regulation. We can in fact use environmental policy as a basis for generalizing about the politics and administration of government regulation in both societies. Analysis of occupational safety and health regulation, consumer protection policy, the regulation of drugs and hazardous substances, and the supervision of financial markets in the two countries reveals that each nation does exhibit a distinctive regulatory style. On the whole, British regulation is relatively informal and flexible while American regulation tends to be more formal and rule-oriented. Britain makes extensive use of self-regulation and encourages close cooperation between governmental officials and representatives of industry. The United States does little of the former and has generally been suspicious of the latter. Both the legislature and the courts consistently play a more active role in making and enforcing regulatory policy in the United States. Yet on balance American workers, consumers, and investors are no better protected than their counterparts in Great Britain.

By controlling nonindustry access to the regulatory process and insulating many of its regulatory bodies from public scrutiny, Britain has attempted—with a fair degree of success—to defuse much of the political conflict associated with government regulation in the United States. Many of Britain's regulatory policies are formulated and implemented through mechanisms of interest-group representation that are essentially corporatist. By contrast, both the making and implementation of government regulation in the United States take place in a large number of highly visible, publicly accessible, and relatively adversarial forums: America's mode of interest-group representation tends to be more pluralist than Britain's.

What is the origin of these national differences in regulatory style? The approaches of the British and American governments to the regulation of corporate social conduct were not always so dissimilar. Government–business relations were highly adversarial during the period of rapid industrial growth in both societies: modest initial efforts to temper some of the worst abuses associated with industrial development met with strong and effective resistance from each nation's industrial community. To read the novels of Charles Dickens or Upton Sinclair is scarcely to be impressed by any substantial differences in the politics or cultures of government regulation in the two societies during their industrial revolutions. The patterns of business–government relations subsequently underwent substantial change in both nations. There are important similarities between the pattern of government regulation of industry established during the 1860s and 1870s in Great Britain—the period of mid-Victorian reform—and that of the Progressive Era in the United States. Both nations established systems of regulation that substituted statutory controls for the common law, provided officials with substantial discretion, made minimal use of prosecution, placed a high value on technical expertise, and encouraged regulatory authorities to act as educators rather than as policemen.

In many respects, the contemporary British approach to regulation resembles the pattern of government regulation adopted in the United States at both the state and federal levels during the Progressive Era. In America, however, the politics and administration of social regulation changed substantially in the late 1960s and early 1970s, becoming more centralized, more legalistic, more visible, and more contentious. This shift can be seen not only in environmental regulation but also in consumer protection and occupational health and safety. Why did one nation respond to its citizenry's increased concern with the externalities associated with industrial growth during the 1960s by making only marginal modifications in the approach to regulation that it had developed

25

a century earlier, while the other chose fundamentally to transform its strategy for controlling the social dimensions of corporate conduct?

In brief, the mid-Victorian style of regulation proved resilient because it rested on three elements: a highly respected civil service, a business community that was prepared to defer to public authority, and a public that was not unduly suspicious of either the motives or the power of industry. Together these three elements make possible a system of regulation based on a high degree of cooperation and trust between industry and government. The legacy of Progressivism, by contrast, proved ephemeral in large measure because America remained very much a "business civilization"—a society in which civil servants continue to enjoy relatively low status, in which business has remained highly mistrustful of government intervention, in which much of the public tends to mistrust both institutions. As a result, while the consultation of industry by regulatory officials continued to be regarded as legitimate in Britain, in America cooperation between industry and government became identified with a betrayal of the public trust.

Each nation's approach to the regulation of industry needs to be understood within the political and social context in which it evolved. For all its myriad inefficiencies, the American approach to social regulation has forced industry to allocate far more resources to environmental and consumer protection and occupational health and safety than it would otherwise have done. Had the American style of regulation remained similar to the British, it is highly unlikely that the United States would have made as much progress as it has over the last two decades in improving its environment and protecting its workers and consumers. On the other hand, had British officials adopted a more aggressive enforcement strategy, they might well have undermined the cooperation of industry on which their system of compliance is based; the result might have been more conflict and only marginally improved compliance. In short, the effectiveness of the British system of regulation rests on securing industry's cooperation to much the same extent that compliance in the United States has required that business be coerced.

At the same time, there is no doubt that the American approach to regulation, for all its well-documented shortcomings, is becoming more widespread. If, during the 1960s and early 1970s, many liberals in the United States viewed the European social democracies, with their highly developed welfare states, their full employment policies, and their labor–management cooperation, as embodiments of an ideal that they hoped America would emulate, then over the last decade much the same could be said of the perception of American government regulation on the part of many people in Europe and Japan. For many environmentalists and consumer and trade union activists throughout the industrialized world, the American style of regulation represents a model to which they would like their nations to move closer. While few appear to believe that government regulation has been more effective in the United States than in their own countries, they are certainly convinced that the effectiveness of their own nation's regulatory system could be enhanced if they had the opportunities and the resources to participate in the regulatory process that their counterparts enjoy in the United States.

While the shortcomings of the adversary relationship have been exhaustively documented by students of government regulation in the United States, the American approach toward regulation developed during the 1970s appears to be the wave of the future while the British seems to be an echo of the past. This perception is particularly evident on the European continent. In a number of respects the politics of pollution control and conservation in the Federal Republic of Germany is similar to that of the United States in the late 1960s and early 1970s. Environmental regulation is now as salient an issue in the FRG as it was in the United States fifteen years ago. In response to public criticism of the adequacy of its controls over emissions from automobiles and coal-burning power plants, the German government has enacted standards that are now the strictest in Europe. Of even greater significance, the rules and regulations emanating from the European Community over the last decade increasingly resemble those adopted by the United States in the early 1970s. Their standards are uniform, relatively strict, and technology-forcing, and they include statutory deadlines; not surprisingly, they have become the focus of growing criticism from the European business community. They have also been the source of considerable tension between Britain and the Community's other members, not only with respect to environmental regulation but also in the areas of consumer protection, corporate governance, and corporate disclosure. Ironically, just as there are some signs that the regulatory process is becoming more cooperative in the United States, it is becoming more adversarial in much of Europe.

Students of American politics have invariably viewed the United States as the most conservative of capitalist polities: its government plays a marginal role in the allocation of capital, it has relatively little public ownership, and its citizens appear to be uniquely committed to the values of capitalism and private enterprise.[3] Yet this portrait of American "exceptionalism" is incapable of accounting for the fact that on balance the United States has attempted to adopt a more coercive approach to the regulation of corporate social conduct than any other industrial society.

The United States remains exceptional, but, at least with respect to this dimension of business–government relations, this exceptionalism is precisely the opposite of what much of the literature on American politics would have led one to expect.

Far from occupying a uniquely privileged position, as Charles Lindblom has described it, industry has been forced to struggle harder to resist additional government restrictions on its prerogatives in the United States than in any other capitalist nation.[4] The United States may be a welfare-state laggard, but over the last two decades it has moved more rapidly and aggressively—though not necessarily more successfully—to tighten controls on corporate social conduct than any other capitalist state. Other nations may have stronger or more radical trade unions, but in no other capitalist polity have middle-class-based public interest movements been so well organized or so influential as in the United States. America may have fewer nationalized firms than any other capitalist nation, but it has nationalized proportionately more land for conservation purposes than any other capitalist polity. (In fact, government ownership of land is more extensive in the United States than in any other capitalist nation.) And while Americans may exhibit a high propensity for risk taking in their business and investment activities, no nation has consistently adopted a more conservative approach to the assessment of the risks associated with new technologies and products than the United States.

This situation is not novel. In *Modern Capitalism*, published in 1965, Andrew Shonfield noted that anyone "coming from Europe and observing the behavior of the people in industry and commerce . . . may well be struck by the way in which it seems to be accepted that it is part of the the lot of businessmen to be pushed around intermittently by one Federal agency or another." He added: "Some government controls over the activities of private enterprise in the United States are unusually fierce, by the standards of other countries." Shonfield specifically cited the Securities and Exchange Commission, which had "established standards for comprehensive and frequent reporting of the affairs of companies . . . far more stringent than anything in Europe." He also observed that "except perhaps for Sweden, labels are nowhere so closely regulated as in the United States."[5] With the emergence of the new wave of social regulation over the last two decades, the contrasts Shonfield noted have become even more pronounced.

The uniqueness of the United States' regulatory style has been amply documented in virtually every comparative study of government regulation. What students of comparative regulatory policy—particularly those in the United States—have overlooked is that Britain's regulatory style is also distinctive. In the area of environmental regulation, for example, Britain's approach contrasts sharply with that of other European nations and of Japan. Britain makes less use of fixed environmental quality or emission standards than any other capitalist nation and relies more extensively on nongovernmental bodies for policy implementation. These differences have become a source of growing tension as the European Economic Community has pressured Britain to harmonize its regulatory policies and procedures with those of the Community's members on the continent. In a number of respects the Community's regulatory policies have more in common with the system of regulation adopted by the United States over the last fifteen years than they have with the British system, which remains rooted in the traditions of the common law.

Virtually every study of Britain written by political scientists and economists over the last decade has emphasized the failures of its government's policies in regard to industry.[6] Compared with those of other capitalist nations, the efforts of the British government to improve the nation's economic performance in the postwar period have been remarkably unsuccessful. The same cannot be said, however, of its efforts to improve corporate social performance. Despite its economic difficulties, the progress Britain has made in preserving and enhancing the quality of its physical environment and in protecting the health and safety of its subjects is roughly comparable to that of many of its more successful competitors. Equally important, its efforts have enjoyed a relatively high degree of legitimacy.

Britain's record certainly compares favorably with that of Japan, whose single-minded commitment to rapid industrial development in the postwar period exposed its citizens to considerable discomfort and injury; by the late 1960s Japan was literally the most polluted nation in the world.[7] It was only after a series of widely publicized catastrophes that the Japanese government first attempted to balance its commitment to economic expansion with the protection of public health and amenity. While the Japanese have devoted considerable resources to improving their physical environment since the special "Pollution Diet" of 1970, controversy continues: violent demonstrations have been held, corporate executives have been publicly humiliated, and the nation has been exposed to a series of protracted, expensive, and extremely bitter lawsuits. The Japanese economy has been able to absorb considerable expenditures on pollution control without impairing its international competitiveness, but the Japanese have found it difficult to integrate environmental concerns and citizen groups into their political system. Likewise, German industry has performed far better than Britain's over the last century, but many of its citizens have recently become extremely dissatisfied with the adequacy of its controls over pollution; hence the emergence of the militant Green party. And while American environmental controls have been relatively effective, its citizens certainly do not perceive them to be so. We have no shortage of explanations for the failures of British industry, but students of British politics have all but ignored Britain's achievements in this other dimension of business–government relations.

It is not only the direction of the differences in government regulation between Great Britain and the United States that is surprising, but their magnitude. Virtually every comparative study of business–government relations has stressed the similarities between the political economies of Great Britain and the United States.[8] In marked contrast to the economic development of both of the nations on the Continent and of Japan, that of the world's first two industrial nations was initiated and managed not by government but by entrepreneurs. As the two classic examples of a "liberal" state, both governments, although now more interventionist than they were during their periods of rapid industrial development, still appear to have less institutional capacity to shape the directions of their nations' economic development than other capitalist polities. Yet, however illuminating this framework for understanding the similarities of economic decision making in Great Britain and the United States, it is clearly unable to account for the radically divergent approaches these two nations have taken to regulating industry.

The study of national regulatory styles thus provides a useful way of exploring the relationship between business and government in both Western Europe and the United States. By linking the study of government regulation to that of comparative politics, it yields important insights into both areas of political science.

# CHAPTER ONE

# The Politics of Environmental Protection in Great Britain

## HISTORICAL BACKGROUND

As the world's first industrial nation, Britain has the world's oldest system of pollution control. The Alkali Inspectorate, established in 1863, was the world's first pollution-control agency, while Britain's conservation movement dates from the latter part of the nineteenth century. Over the ensuing century Britain gradually but steadily expanded its controls over industrial activity: it enacted the world's most extensive system of land-use planning following World War II and significantly strengthened its controls over air pollution in the mid-1950s. Thus before the upsurge of public interest in environmental regulation in the mid-1960s Britain already had in place a fairly extensive system of environmental controls.

The first recorded environmental regulation in the British Isles dates from the thirteenth century. It was promulgated by Edward I, who in 1273 issued a decree prohibiting the burning of "sea coal" in order to protect the health of his subjects. Neither his decree nor a similar one issued by Queen Elizabeth three centuries later, however, led to any discernible improvement in air quality, even though one violator was executed in the sixteenth century. As Britain industrialized in the latter half of the eighteenth century, the increased burning of coal produced dense concentrations of particulates (smoke) in the nation's urban centers. Legislation enacted in 1821 in response to this pollution made it easier for individuals to sue the owners of furnaces that were emitting excessive smoke. This statute did little to prevent the continued deterioration of air quality, however, and the influence of industrial interests in Parliament effectively prevented the passage of any additional pollution-control legislation for the next thirty years.

In 1853 the home secretary, Lord Palmerston, succeeded in persuading Parliament to approve legislation aimed at reducing the amount of smoke emitted by the burning of coal in furnaces in London. Unlike the 1821 statute, which sought to work through the common law, the Smoke Nuisance Abatement (Metropolis) Act specifically declared smoke to be a "nuisance" and provided for penalties for noncompliance. A decade later Parliament enacted its first explicit pollution-control standard. The Alkali Act, approved in 1863, required the manufacturers of alkali—a chemical widely used in the production of soap, glass, and textiles—to remove 95 percent of the hydrochloric acid emitted by their factories. To enforce the standard it established the central government's first pollution-control agency, the Alkali Inspectorate. It consisted of a chief inspector and three associates, all trained in chemical engineering.

In 1881 and again in 1892 Parliament expanded the number of industries that fell under the jurisdiction of the Alkali Inspectorate. In 1906 Parliament approved the Alkali and Works Regulation Act, which listed thirteen "noxious or offensive" gases whose emissions the Inspectorate was responsible for either reducing or dispersing.[1] Subsequent legislation empowered ministers to add to the number of industries regulated by the Inspectorate without the need for additional legislation. This law remained the legal basis for the central government's control of industrial emissions until 1974.

Britain's problems with water quality also date back several centuries. In 1357 King Edward III observed as he rode along the Thames "that dung and other filth had accumulated in divers places upon the banks of the river." He also "noticed the fumes and other abominable stenches arising therefrom." While the Thames remained a good fishing river up to the middle of the eighteenth century, the growth of London, combined with the invention of the water closet, made it into an open sewer. Between 1849 and 1854, 25,000 people died in cholera epidemics, and the death of Prince Albert from typhoid fever in 1861 was attributed to the sewage dumped into the river. In 1885 the scientist Michael Faraday wrote that "the whole of the river was an opaque pale brown fluid . . . near the bridges the feculence rolled up in clouds so dense that they were visible at the surface."[2]

In 1876, Parliament, finding itself unable to conduct its business because of the "noxious odors" coming from the Thames, approved the Rivers Pollution Prevention Act. A member of Parliament remarked in the course of the debate on this legislation: "It is a notorious fact that honourable gentlemen sitting in the committee rooms and the library were utterly unable to remain there in consequence of the stench which arose from the river."[3] Unlike the Alkali Act, which was administered by an agency of the central government, the Rivers Pollution Prevention Act was to be enforced by local authorities. The latter, however, made little effort to implement its provisions, and whatever improvements in water quality took place over the next half century were primarily the result of suits filed by individuals whose private fisheries had been damaged by emissions from industry. Parliament made no further effort to improve the quality of water in Britain until after World War II.

In the last two decades of the nineteenth century, a conservation movement emerged in England. This movement, which was literary as well as political, constituted an integral part of the backlash of the Victorian intelligentsia against the values of economic liberalism. Its early leadership included many prominent social philosophers, among them John Ruskin, John Stuart Mill, and William Morris. The orientation of this movement was, literally, a conservative one: its adherents sought to conserve the values of traditional England—its unspoiled countryside, its ancient buildings, its underdeveloped wilderness, and its wildlife—from the "progress" of industrial civilization.

The British upper and upper-middle classes came to regard the English countryside in much the same way that many Americans during the Progressive Era began to view the wilderness. In each case, preserving undeveloped land (and in the English case buildings as well) from commercial exploitation was a way for the nation to preserve not simply a vital part of its heritage but its moral and social character. As Morris argued, when we destroy a beautiful old building, "we are destroying the pleasure, the culture, in a word, the humanity of unborn generations." Patrick Abercrombie subsequently remarked that "the greatest historical monument that we possess, the most essential thing which *is* England, is the countryside, the market town, the village, the Hedgerow trees, the lands, the Capes, the streams and the farmlands." Like the conservation movement of the Progressive Era, the British movement of the late nineteenth century was elitist: "Victorian preservationism was distinctly a gentlemanly avocation pursued by cultured people well removed from, and indeed averse to, the base pursuits of trade and manufacture."[4]

The origins of the contemporary British environmental movement date from this period. The earliest recorded environmental organization was the Commons Preservation Society, established "to restore to the Commons something of the attributes of the ancient Saxon Folk-land."[5] In 1889 a group of residents of Manchester established a Fur and Feather Group to protest against the use of birds' feathers in the making of hats.

At about the same time, several members of the English gentry, upset by the threat to open spaces posed by industrial and urban expansion, founded the National Trust for Places of Historic Interest and Natural Beauty. Its objective was to acquire land and buildings by either gift or purchase in order to protect them from economic development and thereby to ensure both their preservation and public access to them. The Trust saw itself not simply as a landowning body but as "the national champion of the preservationist cause." Its first leaflet declared:

> The National Trust is not only a holder of natural scenery and ancient buildings, but it also does what it can to promote local interest in the preservation of any worthy historical object or natural beauty. Whether it be a waterfall destroyed ... or the need of obtaining ... a pleasure ground ... for the people, the Trust, working sometimes alone, at other times in conjunction with kindred societies, brings its influence to bear in the direction and spirit of its promoters.[6]

In 1907 the Trust was reconstituted as a statutory body by an act of Parliament and charged with the "permanent preservation" of property "for the benefit of the nation."[7] It was also given the authority to declare its land and buildings inalienable, so that the approval of Parliament would be required to change their status. During the same decade the Fur and Feathers Group received a royal charter and became the Royal Society for the Protection of Birds.

The interwar years in Britain witnessed a substantial increase in public concern about amenity issues. During the 1920s the Ancient Monuments Society, the Ramblers' Association, the Council for the Preservation of Rural England and its Scottish and Welsh equivalents, the National Trust for Scotland, and the Central Council for River Protection were established. The formation of the Ramblers' Association, whose purpose was to secure access to the countryside for hikers, was the first British environmental organization whose membership was drawn primarily from the working class. The most important organizations established during this period were the Councils for the Preservation of Rural England, Scotland, and Wales. Formed in large measure as a response to the threats to the countryside posed by urban sprawl and the introduction of the automobile, they initially concentrated on restricting ribbon development and roadside advertisements.

## THE POSTWAR PERIOD

The two decades following the end of World War II saw a significant expansion of the British government's efforts to reduce pollution and protect the physical environment. Among the most important legislative accomplishments of the postwar Labour government was the Town and Country Planning Act. Approved in 1947, it represented, in the words of one observer, "probably the single most important legislation [Britain] had ever seen."[8] Parliament had twice previously attempted to regulate land use: a Housing, Town Planning, etc. Act had been approved in 1909 and a more comprehensive Town and Country Planning Act was enacted in 1932. Their impact, however, had been limited by both a shortage of funds and the administrative capacity of the local authorities responsible for implementing them. Following World War II the government came under strong pressure to establish an effective system of land-use planning, in part as a way of compensating the public for the physical destruction the country had experienced. As a government document put it: "There must be some plan of action ready to reward the valiant; works that can be put into immediate operation and will later fall into their ultimate place."[9] The evident success of British wartime planning also strengthened public confidence in the potential effectiveness of planning by government.

The Town and Country Planning Act of 1947 and a companion measure enacted for Scotland fundamentally transformed the administration of land-use planning in Britain. In effect, the right to use land was nationalized. Each county council and county borough was designated a local planning authority, and no private development could take place without their consent. All development rights for underdeveloped land were transferred to the central government, which set aside £300 million to compensate its owners. In addition, local planning authorities were given the power to require the demolition of existing buildings, preserve woodlands, protect buildings of architectural or historical interest, control outdoor advertisements, and require the proper maintenance of wastelands. They were also required to establish development plans that described the intended use of the land within their jurisdiction.

This legislation played an important role in encouraging the formation of local organizations with an interest in planning and conservation. In fact, "local environmental groups appear to have arisen largely in response to the postwar planning system and the powers and responsibilities it conferred on local authorities."[10] The first local amenity society was established in 1846; the rapid growth of such groups during the 1950s was due primarily to the emergence of specific controversies over the development decisions made by local authorities under the provisions of the 1947 legislation.

The other important land-use statute enacted by Clement Attlee's government was the National Parks and Access to the Countryside Act of 1949. This law officially established for the first time a system of national parks in Britain. A National Parks Commission (now called the Countryside Commission) was established to select areas for designation as national parks, although its recommendations required the approval of the minister of Housing and Local Government (now the Department of the Environment). By 1957 the commission had designated a total of ten parks, totaling more than 5,251 square miles. No central agency was established to administer the parklands, however, and the ownership of both surface and mineral rights remained in private hands. Nor was commercial development prohibited, provided proper safeguards were followed. Lewis Silkin, the minister of Town and Country Planning, informed Parliament that mining would be permitted in the national parks under three conditions:

> It must be demonstrated quite clearly that the exploitation of those minerals is absolutely necessary in the public interest. It must be clear beyond all possible doubt that there is no possible alternative source of supply, and if those two conditions are satisfied, then the permission must be subject to the condition that restoration takes place at the earliest possible opportunity.[11]

Moreover, the responsibility for administering the parks remained primarily in the hands of local planning authorities, many of whom remained reluctant to impose restrictions on commercial activity that would limit their tax base or reduce employment. In essence, the 1949 legislation amounted to more of a moral than a legal or financial commitment on the part of the government to encourage the preservation of scenic areas and promote their use for open-air recreation.

Another section of this statute authorized the National Parks Commission to recommend the designation of "areas of outstanding beauty," although again no legal protection was extended to the areas selected and no funds were appropriated to help preserve or administer them. The role of the National Parks Commission was primarily advisory. As a statutory authority, it was required to be consulted before other governmental bodies, local as well as national, made decisions affecting areas for which it was legally responsible. While its recommendations were not always followed, it did serve to legitimate the importance of conservation in the making of policy by both governmental and quasi-governmental units.

Third, the 1949 legislation established a Nature Conservancy (now called the Nature Conservancy Council) to establish, maintain, and manage Britain's nature reserves. The Conservancy was provided with a professional conservation and research staff to enable it to establish a national system of nature reserves and empowered to acquire land compulsorily if it was unable to reach agreement with its owner. By the mid-1970s it was responsible for maintaining and managing 167 national nature reserves, totaling 328,000 acres.

Finally, a major accomplishment of the 1949 legislation was to facilitate public access to the countryside—a cause for which the Commons, Open Spaces, and Footpaths Preservation Society had begun campaigning in 1865. In their history of the national parks, Ann and Malcolm MacEwen write:

> One of the more successful parts of the 1949 Act, Part IV on public rights of way, broke entirely new ground by giving the public statutory rights it had never enjoyed before. Almost at a stroke it arrested the piecemeal and widespread loss of rights of way that had been taking place throughout the 1930s and 1940s and earlier—a process that could be arrested previously only by taking costly and risky legal proceedings for each individual path separately.[12]

Underlying this provision was

> a feeling that the wilder parts of Great Britain were the possession of all the people, and not merely of the landlord who owned the title deeds of the grouse-moor; a feeling, too, that the mountains and forests were a place where healthy exercise could be obtained by those who lived in the cramped back streets of the industrial towns.[13]

The Conservative governments that held office between 1952 and 1964 continued to expand public controls over both industrial emissions and land use. During this period Parliament enacted the Navigable Waters Act (1953), the Protection of Birds Act (1954), the Rural Water Supplies and Sewage Act (1955), the Litter Act (1958), the Radioactive Substances Act (1960), the Estuaries and Tidal Waters Act (1960), the Rivers (Prevention of Pollution) Act (1961), the Deer Act (1963), and the Water Resources Act (1963). The most important single change in British environmental policy during the 1950s, however, involved the control of air pollution.

The smoke produced by the burning of coal in both homes and factories had long been regarded as Britain's most serious environmental problem: it not only severely restricted visibility (hence London's famous fogs) but constituted a clear threat to public health. While "black" smoke was declared a nuisance by the Public Health Act of 1875, local authorities made little effort to enforce this statute. In 1889, 5,800 citizens of

*37*

the industrial city of Sheffield signed a petition calling for the prosecution of industrial smoke polluters, but it had little effect. Following a particularly dark fog over London in 1898, the Coal Smoke Abatement Society was formed. Its objectives were twofold: to pressure the city of London into enforcing its own smoke-control ordinances and to demand that Parliament permit the city to enact more effective controls. After more than a quarter century of agitation, Parliament finally enacted the Public Health (Smoke Abatement) Act of 1926. The legislation, however, was drafted in such a way as to make its restrictions on coal burning virtually impossible to enforce. Parliament simply remained unwilling to require householders to switch to the burning of smokeless fuels; indeed, the Public Health Act of 1936 specifically forbade local authorities to prosecute householders for smoke emissions. Not only might such a policy hurt the domestic coal industry—already severely depressed—but it would interfere with what had become an important English tradition, the "pokeable, companionable fire."[14]

In 1946 the city of Manchester had managed, through a private act of Parliament, to obtain the right to establish "smokeless" zones within its jurisdiction. Over the next decade nineteen additional local authorities obtained similar powers. But the central government, in its "role as protector of individual freedom against [the] encroachments [of local governments]," continued to resist any expansion of the latter's power over their citizens.[15] The political climate, however, changed dramatically in the early 1950s. On December 5, 1952, an "unusually nasty" fog descended over London, virtually eliminating visibility for nearly four days.[16] The total number of deaths ultimately attributable to this inversion, which was produced primarily by the household burning of coal, was estimated at 4,000.

In May 1953 the government agreed to establish a committee of inquiry. The Beaver Committee's final report, issued six months later, stated emphatically that "air pollution on the scale with which we are familiar in this country today is a social and economic evil which should no longer be tolerated." It recommended that the government prohibit emissions of dark smoke from chimneys, require that industrial plants that burned solid fuel above a certain volume install equipment to arrest grit and dust, and most important, allow local authorities—subject to ministerial approval—to establish smokeless zones and smoke-control areas. It also urged a significant expansion of the jurisdiction of the Alkali Inspectorate to make it responsible for "controlling all emissions from industrial processes, including smoke, grit, and dust, where abatement presented special difficulties."[17] The cost of implementing these recommendations was estimated by the committee at approximately £250 million a year.

Even before the committee had released its report, the City of London had proposed legislation giving it the authority to create smokeless zones within its jurisdiction. Parliament approved the city's request and shortly afterward, on July 5, 1956, a Clean Air Bill, embodying most of the Beaver Committee's recommendations, received the royal assent. Its most important provision gave local authorities the power to declare smokeless zones within their jurisdictions—a policy that had been urged by the National Society for Clean Air for nearly thirty years. Since smokeless fuels (e.g., types of coke) could not be burned in open grates—the standard method of home heating—the act also provided for government grants to households to help defray the costs of purchasing and installing new heating systems. Equally significant, the passage of the Clean Air Act marked "the beginning of a period of greater public and political concern which ensured a higher priority for the control of air pollution" on the part of local public health officials, who formerly had devoted very little attention to smoke abatement.[18] Two years later a second key recommendation of the Beaver Committee was adopted: the number of plants under the jurisdiction of the Alkali Inspectorate was more than doubled, so that the locus of responsibility for controlling most complex industrial sources of air pollution was effectively shifted from local authorities to the central government.

CONTEMPORARY REEMERGENCE

Public interest in environmental issues in Britain increased substantially during the second half of the 1960s. The British public was influenced by and in turn influenced the dramatic explosion of concern about the state of the environment that occurred simultaneously in every industrial nation—a convergence symbolized by the United Nations Conference on the Environment held in Stockholm in 1972. Beginning in the mid-1960s, environmental pressure groups became both better organized and more politically active throughout Western Europe, Japan, and the United States. Between 1970 and 1975, nearly every member nation of the Organization for Economic Cooperation and Development (OECD) enacted additional environmental legislation,[19] and in 1973 the European Economic Community approved the first in a series of action plans designed to strengthen environmental controls in its member states.

Britain, like many other industrial nations, experienced several natural disasters during this period. In October 1966 an avalanche of coal sludge buried 28 adults and 116 children in the Welsh village of Aberfan. The disaster stunned the nation and a subsequent national inquiry severely criticized the National Coal Board. In his book *Derelict Britain*, John Barr characterized the mountains of coal sludge ringing many of Britain's collieries as "monuments to man's degrading presence."[20] Stanley Johnson subsequently noted that "Aberfan was important not because it had a clear place in the genealogy of pollution but because it prepared the way for a greater understanding of what pollution was. Consciously or unconsciously, the Aberfan disaster changed man's thinking."[21]

A few months later a giant oil tanker, *Torrey Canyon*, crashed onto the rocks off the coast of Cornwall, spilling more than 60,000 tons of crude oil into the Atlantic and onto the beaches of England and France. The biggest single pollution disaster ever recorded, the *Torrey Canyon* wreck had a profound effect on public awareness of pollution problems in Britain. It also received extensive coverage in the world press. The incident "caused much public alarm and brought considerable criticism of the Government for its lack of preparedness." Harold Wilson's government responded by establishing an ad hoc committee of scientists under the chairmanship of Lord Solly Zuckerman and referred the "question of future measures against the pollution of our shores to the Select Committee on Science and Technology of the House of Commons."[22] Both committees subsequently released reports highly critical of the administrative and scientific capacity of the government to deal with pollution problems.

Rachel Carson's best-selling *Silent Spring*, published in the United States in 1962, attracted considerable attention in Britain. Two years later, the annual report of the Nature Conservancy revealed that all of the fifty-two eggs of eleven species of seabirds collected at four sites around the British coast contained residues of organochloride compounds. A year later, a joint committee of the British Trust for Ornithology and the Royal Society for the Protection of Birds reported that 96 percent of the 236 corpses of birds it examined contained organochloride pesticides. In the fall of 1969, between 50,000 and 100,000 seabirds were found dead along the British coastline, the victims of the dumping of the toxic chemical PCB into rivers and directly into the ocean itself. This greatest seabird disaster in British history provoked considerable public concern over the effects of the indiscriminate dumping of industrial wastes.

All of these developments contributed to a substantial increase in the membership of environmental organizations in Great Britain. Between 1967 and 1980, the membership of the Council for the Protection of

Rural England (CPRE; formerly the Council for the Preservation of Rural England) increased from 14,000 to 31,000; that of the National Trust from 159,000 to 1 million; that of the Royal Society for the Protection of Birds from 38,000 to 300,000; that of the Ramblers' Association from 16,000 to 32,000; and that of the Society for the Promotion of Nature Conservation from 29,000 to 129,000.[23] In 1958 local amenity societies had been established in 200 British communities; by 1975 their number had increased more than sixfold. In 1977 the total membership of the local amenity movement was estimated at 300,000, or slightly less than 1 percent of the British population.[24] Allowing for overlapping membership, by the middle of the 1970s between 2.5 and 3 million British citizens belonged to at least one environmental organization. In addition, three new national environmental organizations were founded: Greenpeace, a British chapter of the American organization Friends of the Earth (FOE), and the Conservation Society. In its first year of existence the Conservation Society recruited 8,000 subscribers, while within four years of its founding FOE had established more than a hundred local chapters.

During this period the attention devoted by the media to environmental issues also increased substantially. While the coverage of environmental issues in the *Times* was roughly stable between 1953 and 1965, it increased by 182 percent between 1965 and 1969 and by 281 percent between 1965 and 1973. The space devoted to letters on environmental issues in the *Times*, generally "considered an influential forum for the discussion of major issues," expanded from 144 square centimeters in 1965 to 3,305 square centimeters in 1973.[25] At the same time, the definition of what constituted an environmental issue was broadened considerably. During the 1950s, coverage of the environment focused almost exclusively on the landscape and roads. By the mid-1960s these two issues constituted only a minority of the articles on the environment featured in the *Times*. They had become supplemented by articles on road transport, international environmental developments, water pollution, resource use, air transportation, air pollution, and environmental organizations themselves.

Up through the 1960s, many British environmental organizations were extremely reluctant to encourage publicity, preferring instead to rely on private negotiations with government officials. But over the following two decades environmental groups came to rely increasingly on the shaping of public opinion as a way of influencing public policy. Commenting on some of the changes that occurred following the retirement of the organization's general secretary in 1965, a senior staff member of the CPRE noted:

Under [Sir Herbert] Griffin, publicity was very restrained. The CPRE did not seek attention. "Do good by stealth and be found out by accident" was his motto. His style of operation was through personal contact in the corridors of power. He fastidiously avoided embarrassing those whom he influenced or sought to influence.... Nowadays we are not reluctant to go public. Indeed, we are very media-conscious. This is better suited to the general style of environmental politics which has become more conflict-oriented.[26]

Over the last two decades several campaigns have been launched to stimulate public awareness of environmental issues. These efforts have included the "Countryside in 1970" conferences held in 1963, 1965, and 1970, the European Conservation Year (1970), and the European Architectural Heritage Year (1975). In 1972 a group of journalists established an Environmental Communicators' Organization to encourage the press to devote more attention to environmental problems. One observer noted, "The preoccupation of the communications media with the 'environment' transformed hitherto 'low visibility' decisions into highly conspicuous and therefore palpable indications of a minister's personal scale of values."[27]

Increased interest in environmental issues was also reflected in the deliberations of Parliament. Between October 1970 and December 1971, members of the House of Commons tabled to all departments an estimated 1,579 questions on environmental matters, while approximately twenty early-day motions expressing concern over such matters as land use, noise from jet flights, and heavy metal pollution were tabled.[28] By 1973 several permanent all-party committees had been established to discuss environmental issues, including the Conservation Group (concerned with historic buildings), the Ecology Group, the Anti-Pollution Group, the Select Committee on the Environment, the Inland Waterway Group, and the Committee for the Conservation of Species and Habitats. Peers and MPs introduced private members' bills dealing with specific environmental issues, and both houses of Parliament now include members who work closely with various environmental organizations.

A number of MPs hold official or honorary positions in many [environmental] organizations, and others are undoubtedly sympathetic to the environmental cause. Through the use of the backbenchers' traditional weapons—posing Parliamentary Questions, sponsoring motions and moving adjournment debates—sympathetic members can impress upon ministers the weight of the environmentalist cause, and elicit their support.[29]

## The Political Response

In 1969 Harold Wilson announced in the House of Commons that the Queen had approved his recommendation to establish a Standing (i.e., permanent) Royal Commission on Environmental Pollution. Wilson also established a permanent Central Scientific Unit on Pollution within the Ministry of Housing and Local Government to help coordinate the government's pollution-control efforts. Shortly before the 1970 British general election the Wilson government hurriedly issued a white paper titled "Protection of the Environment: The Fight against Pollution," which summarized and reviewed existing programs dealing with pollution control and promulgated several new policies. While the environment played a negligible role in the 1970 British general election, it had become an established part of the political agenda. Writing in 1972 Roy Gregory remarked: "The environmental lobby has succeeded in creating a 'halo effect' for the conservationist cause; today it is a bold politician who risks its disapprobation. The signs are that this network of pressure groups has now become a feature of lasting importance on the British political scene."[30] A year later, William Solesbury observed that "the environment as a general issue has risen to a dominant position among political concerns."[31]

Following the election of a Conservative government in 1970, the word "environment" appeared for the first time in an address by the Queen. Her Majesty promised that her ministers would "intensify the drive to remedy past damage to the environment and . . . seek to safeguard the beauty of the British countryside and the seashore for the future." Shortly afterward Edward Heath's government issued a major white paper on the reorganization of central government, which announced, among other items, the establishment of a Department of Environment (DOE). This department assumed a series of wide-ranging administrative and statutory powers over a "whole range of functions which affect people's environment," including land-use planning, surface transportation, housing construction, the preservation of amenity, the protection of the coasts and countryside, the preservation of historic towns and monuments, and the control of air, water, and noise pollution.[32] One journalist wrote:

> The Department of the Environment was, in fact, a very novel creation. For the first time in England it concentrated under one minister as well as in one ministry all the statutory powers, patronage, budgetary control, and political decision-making that had previously belonged to three separate and influential ministers and ministries: Transport, Public Building and Works and Housing and Local Government.[33]

Measured in terms of the number of white-collar employees, it became the third largest department in the British government.

In 1971 the DOE's first secretary of state, Peter Walker, proposed legislation to reorganize the more than 1,400 local and regional agencies responsible for water supply, sewerage, and water conservation in England and Wales. In their place he planned to establish ten regional water authorities. Their boundaries would be based on natural watersheds rather than on the jurisdictions of local governmental units. Each would be responsible for all policies having to do with water, "literally from the source to the tap."[34] The impetus for this proposal came from officials of the Ministry of Housing and Local Government (one of the predecessor bodies of the DOE), who felt that local authorities were neglecting their responsibilities for water management.

After lengthy and intense negotiations—more than a hundred organizations submitted written evidence to the Central Advisory Water Committee—legislation was approved by Parliament in 1973 and implemented the following year. The Water Act (1973) represented the most far-reaching reform in the history of British water administration. Indeed, it has been called "as radical a policy switch as can be cited in postwar Britain."[35] While environmental pressure groups did not play a critical role in shaping this legislation, one of its immediate effects was to "greatly strengthen the pollution controller's hand."[36]

A year later Parliament enacted the Control of Pollution Act. This legislation was originally proposed by the Conservative government but was enacted under the newly elected Labour government. Unlike previous such statutes, each of which had addressed a specific form of pollution, this act addressed a multiplicity of environmental problems, including waste disposal, water pollution, noise nuisance, and air pollution. It thus marked "the first formal recognition of the environment as a single entity," and was regarded as "a major step forward in the administration of pollution control in the UK."[37] While the law primarily represented a consolidation of previously enacted statutes, it did include some important innovations. One provision granted the secretary of state the power to regulate the composition of oil used in motor vehicles and furnaces. Another gave district councils, for the first time, the authority to collect and publish information on air pollution in their districts. Each water authority was required to maintain a register containing details of all applications for permission to discharge effluents, the terms under which such applications were approved, and any subsequent samples and analyses of effluents, and to open this register to public inspection. In both cases the Control of Pollution Act specifically reversed the policies established by previous legislation, which had placed severe restrictions on the ability of pollution-control agencies to release information they obtained from polluters to the public.

The reorganization of local government, which also took place in 1974, substantially improved the effectiveness of environmental regulation at the local level. By amalgamating many smaller local authorities, it created environmental health departments that were far better equipped to handle many pollution problems than their predecessors had been. That same year, Parliament enacted the Dumping at Sea Act, which replaced a system of voluntary controls over dumping in British waters and from British ships in external waters with a system of statutory controls. The British government also approved a number of new policies designed to protect plants and animals. The Conservation of Wild Creatures and Wild Plants Act (1975) extended protection to a small group of endangered animals, somewhat more limited protection to all wild plants, and more specific protection to a short list of rare species; provision was made for additional species to be added by the secretary of state for the environment on the advice of the Nature Conservancy Council. Britain also signed the 1972 Washington Convention on International Trade in Endangered Species and supported the European Community's ban on the import of primary products made from whales. The Endangered Species (Import and Export) Act, enacted in 1976 primarily as a response to vigorous pressure from the Friends of the Earth, placed restrictions on international trade in a number of endangered species. The British government also played a leading role in securing Community adoption of a directive banning the slaughter of wild birds.

## THE CONTEMPORARY BRITISH ENVIRONMENTAL MOVEMENT

The membership of British environmental organizations is more affluent and better educated than the population as a whole. In Britain this division has a geographical dimension as well: amenity organizations are more active in the south of England, which contains much of Britain's white-collar and professional community, than they are in the industrial Midlands and in the north. Many community residents have become active in environmental organizations, in part as a way of preventing additional economic development that threatens to undermine the value of their property and their semirural lifestyle. As a result, many local environmental conflicts have pitted members of the working class, dependent on industry for employment, against members of the middle class, who tend to have a greater stake in the preservation of amenity values, particularly when they are not dependent on the community in which they reside as a source of employment.

On the whole, the British trade union movement has not been active in the shaping of environmental policy. It has tended to become involved only when the economic interests of its members have been directly at stake. The Trades Union Council supported increased public subsidy for rail transportation, for example, in order to protect the jobs of its members. Trade unions have supported coal mining in the Vale of Belvoir—strongly opposed by British amenity organizations—and local unions have frequently backed industry's efforts to secure approval for investments that promise to create employment, often at the expense of environmental quality. Unions have been divided on the issue of nuclear power: those representing electrical workers support it and those representing coal miners oppose it.

While the social base of British environmental organizations has expanded considerably since the nineteenth century, members of the aristocracy continue to play an important role in the British environmental movement. As a kind of British counterpart to Earth Day, in 1970 the Duke of Edinburgh (Prince Philip) formally inaugurated a movement called "The Countryside in 1970," which sought to focus national attention on the state of the British countryside. When the National Trust launched an appeal for funds to purchase 980 miles of the coast of England, Wales, and Northern Ireland in order to preserve public access to it, this campaign—dubbed Operation Neptune—was headed by Prince Philip. More recently, Prince Charles has become identified with a variety of conservation issues.

Many leaders of the older, more traditional environmental organizations, such as the Royal Society for the Protection of Birds and the National Trust, have the same social background as the Victorian aristocrats who originally established them. The National Trust in particular retains extremely close ties with "the landed interests which in turn support the Trust and exercise great influence within it."[38] Not coincidentally, the Trust remains the most "established" and respectable British environmental organization, enjoying considerable influence and a high degree of public support.

In part for this reason, much British environmental legislation, particularly in the areas of conservation policy and plant and wildlife protection, has been either initiated or primarily debated in the House of Lords. In addition, peers have frequently offered important amendments to environmental legislation approved by the House of Commons. A committee of the Lords with particular expertise in environmental matters—the Environmental Sub-committee of the Special Committee on the European Communities—has played an important role both in shaping British environmental policy and in representing British interests within the European Community. Two of its members, Lord Ashby and Baroness White, have also served on the Royal Commission on Environmental Pollution; the former was its first chairman.

Many hereditary peers are large landowners and have a personal interest in many aspects of rural conservation and the preservation of historic buildings. Some are active in a wide variety of nongovernmental and quasi-governmental environmental organizations. Lord Creighton, for example, in 1981 was chairman of the Council of Environmental Conservation, the Zoo Federation of Great Britain, the Fauna and Flora Preservation Society, and the All-Party Conservation Committee, and vice-president of the Wild Life Fund (UK). On the other hand, the commitment of the British landed aristocracy to environmental protection is selective: while frequently critical of the environmental effects of industrial activity, they have been less sympathetic to the complaints of many British environmentalists about the destruction of the countryside by commercial farming interests. Much the same can be said of the role of the National Farmers Union: it has strongly supported strict controls over industrial emissions and siting but vigorously resisted the efforts of amenity organizations to impose planning controls over agricultural land use.

If the National Trust is the most "respectable" British environmental organization, the Friends of the Earth has been one of the more controversial. It is the British environmental group whose political style most closely resembles that of environmental groups in the United States. FOE has been unique in its media-oriented and confrontational style of politics: its political debut consisted of dumping hundreds of empty nonreturnable bottles on the doorsteps of the corporate headquarters of Schweppes in order to dramatize the need for restrictions on such bottles. This move represented

> a significant departure from traditional environmental lobbying in Britain characterized by low-keyed representations through established channels. ... It also incorporated notions of participatory democracy and forms of direct political protest such as boycotts, sit-ins, marches and demonstrations borrowed from the student movement.[39]

One journalist wrote in the *New Statesman*:

> Without ever flouting the law, FOE gave Britain's honourable but cautious environmental lobby, with its concern for hedgerows, footpaths and vanishing peregrines, a much-needed radical edge. They relished confrontation and embarked on a programme of news-grabbing gimmickry. They

launched a paper chase and built 40 paper mountains to show how money could be saved by recycling paper. They haunted the international Whaling Commission with inflatable plastic whales . . . and sent the Packaging Council a 30-foot sack of rubbish as a Christmas present.[40]

From May 1971 through April 1980, FOE organized a total of thirty-three national demonstrations; a third of these activities protested the killing of whales, while others dealt with such issues as throwaway bottles, nuclear power, and transportation policy. Many of these demonstrations succeeded in capturing considerable media attention; FOE's most important political resource has been its skill in generating publicity. Its most important political accomplishments have included the passage of the Endangered Species Act of 1976 and the agreement of the British government to hold a public inquiry into a proposal to process nuclear waste at Windscale. By 1980 it had approximately 15,000 members organized into 250 local chapters and was able to support a staff of 24 at its national office in London.

While most lobbying by environmental groups has taken place through "normal" political channels, the movement also has reflected the traditional strand in British radicalism of nonviolent civil disobedience. Individuals have on occasion blocked entrances to factories and construction sites, immobilized heavy lorries, and disrupted motorway inquiries by noisy protests and sit-ins. During the mid-1970s, efforts were launched to physically block the construction of the Torness nuclear power station, and in the early 1980s scuba divers from Greenpeace attempted to plug the pipeline through which the Windscale reprocessing plant discharged radioactive water into the sea. In general, however, the British environmental movement has been both nonviolent and law-abiding.

Except for the rather insignificant Ecology party, the British environmental movement has not itself campaigned directly for public office.[41] Nor has it become identified with either the Labour or Conservative party. Each party encompasses both strong supporters and opponents of amenity values. While some of Labour's middle-class activists are supporters of strong environmental controls on industry—in the early 1970s they formed a small group called the Socialist Environment and Resources Association—the trade union constituents of the party have generally opposed environmental regulations that threaten the employment of their members. Some local authorities controlled by the Labour party, however, have enforced relatively strict controls over smoke emissions. While many of the Conservative party's supporters in Britain's middle-class suburbs are very committed to amenity values, other constituencies of the party, including industrialists and large landowners, often find themselves at loggerheads with amenity groups. Traditionally, Britain's "greenist" party has been the Liberal party. The recently established Social Democratic party does draw a disproportionate amount of its support from the same social base as many environmental organizations, but, at least in the first general election it fought, it did not make environmental protection an important campaign issue.

With the exception of the controversy over the banning of lead in petrol, disputes over British environmental policy have not been fought on partisan lines. British environmental policy has exhibited considerable stability over the last fifteen years: the four changes in government that have taken place in Britain since 1970 have had no discernible effect on its direction. Nor has environmental regulation been an important issue in any British general election, though since the early 1970s most party manifestos have included some reference to it. This lack of partisan conflict over environmental policy is the other reason that the Lords has played a more important role than the Commons in shaping public policy in this area.

## The Environmental Movement and Government

A unique feature of British environmental policy is the extent to which functions that are carried out primarily by governmental bodies in other industrial countries are in Britain performed by private bodies. The National Trust, for example, rather than the British government, acquires historic buildings and scenic land areas to preserve them for public use. (The Trust does not normally purchase land, but usually receives it as a gift.) The government supports the Trust directly through grants and indirectly through regulations governing its tax status. The Trust also receives considerable funds from the public, including the dues paid by its million members. As a result, the Trust has become the largest private landowner in England and Wales, owning more than 1 percent of the land surfaces of these two countries and more than 200 historic buildings within them. A similar role is performed by the Royal Society for Nature Conservation; working in close cooperation with the Nature Conservancy Council—from which it also receives financial support—the society either owns or leases more than 100,000 acres of nature reserves. The Royal Society for the Protection of Birds, whose 320,000 subscribing members make it the largest wildlife conservation body in Europe, either owns or administers seventy-seven reserves of national or international importance throughout Britain.

The British government is in fact a major source of financial support

*NT largest landowner in UK*

for many environmental organizations; in one survey of British environmental groups, more than one-third listed the British government as among their three most important sources of funds; indeed, for 23 percent it was either their first or second source of income.[42] Some environmental organizations, such as the Building Conservation Trust and the Societies for the Preservation (now Protection) of Scotland, England, and Wales, were actually established by the government, while a grant from the Countryside Commission resulted in a complete restructuring of the Standing Committee on National Parks. In addition, the government frequently relies on environmental organizations to assist it in policy implementation. The Keep Britain Tidy Group, for example, is the "chosen instrument of the Government for carrying out its policy in the field of litter prevention and abatement" and receives more than half of its income in the form of a grant from the DOE.[43] The Nature Conservancy relies on the Royal Society for Nature Conservation to help it administer the sites of special scientific interest (SSSI) for which it is responsible. While in every other European country the program for the European Architectural Heritage year was sponsored by the government, in Britain it was coordinated by the Civic Trust with financial assistance from the Treasury.

British environmental groups clearly benefit from the widespread pattern of consultation and advice sought by governmental bodies before decisions are taken; in fact, nearly half of the British environmental organizations have representatives on one or more official advisory committees. Some environmental organizations enjoy official or quasi-official status. The Council for the Protection of Rural England—which is actually a representative body made up of fifty-four constituent organizations, including various professional associations, monitoring associations, local authority associations, and many of the major national amenity and preservation societies—has achieved a semiofficial watchdog status in relation to the English planning system. The Civic Trust and local amenity societies enjoy a similar status vis-à-vis policies affecting conservation areas. The Ramblers' Association is regularly consulted over issues involving access to the countryside, and the Council for the Protection of Rural Scotland works closely with the Scottish Development Department.

Under the Town and Country Planning Act of 1968, local planning authorities must consult specified voluntary organizations before allowing a listed building to be altered or demolished, and the Control of Pollution Act of 1974 obligates local authorities to consult the Keep Britain Tidy Group in designing plans for litter abatement. More informally, many local planning authorities rely heavily on the expertise of various amenity groups in both evaluating and challenging various development projects. A revolving door exists between various environmental organizations and the staffs of governmental units with important environmental responsibilities, such as the Countryside Commission, the Historic Buildings Council, the National Water Council, and the Nature Conservancy Council. Such government agencies frequently draw on the political resources and technical expertise of environmental groups to help them mobilize support in the intergovernmental competition for administrative responsibilities, legislative time, and funds.

By the late 1960s there was widespread consensus among environmentalists, the government, planning officials, and industry that it would be useful to have one body to coordinate the activities of the entire environmental movement and represent its views to the government. To this end the Committee for Environmental Conservation was established in 1969. (In 1980 it assumed the function of the Council for Nature and its name was changed to the Council for Environmental Conservation [CoEnCo].) It is comprised of nominated representatives from each of the main environmental organizations, with other environmental organizations represented on its various subcommittees. CoEnCo's purpose is to "speak with authority on all aspects of national conservation policy."[44] The government has invited it to submit evidence on the management and conservation of water resources, the manufacture and disposal of metal containers, the amenity use of water space, and controls for mining and planning. CoEnCo is financially supported by various foundations, donations from industry and government, and contributions by its member organizations. Recently the Confederation of British Industries has begun to send representatives to its meetings.

Given the relatively small role played by Parliament in the shaping of British environmental policy, the most effective means by which an environmental organization can affect national policy is to enjoy a close working relationship with a particular governmental agency. When such a relationship exists, the organization can expect to be consulted regularly by civil servants when they are considering new policy initiatives, and its leadership is likely to be appointed to various advisory bodies or royal commissions formed to make policy recommendations. These privileges, however, come with a price attached: they are given only to organizations that are considered "respectable" and whose leaders can be counted upon to behave "responsibly." As a result, the process of consultation also serves as a means of co-optation.

The exercise of influence is a two-way process. Groups grow into elaborate consultative procedures with government are induced to moderate their demands and tactics. Much time can be spent responding to a flow of consultative documents from government, sitting on official committees and providing information for policy making. Group leaders become enveloped in the consensual atmosphere of Whitehall with civil servants attempting to explain the constraints on government action and the rival claims that have to be balanced.[45]

The experience of Friends of the Earth, originally the most militant and activist wing of the British environmental movement, furnishes a graphic illustration of this process. By its tenth anniversary its leaders found themselves devoting much of their energies to serving on various DOE working bodies and providing technical advice to civil servants. Thus following the dumping of bottles in front of Schweppes, they became members of a DOE advisory committee formed to consider new strategies for litter abatement. Revealingly, an article published in 1974 by the Conservation Society, the other militant environmental group formed in the late 1960s, cautioned: "The Society must not be, or seen to be, merely a vehicle for anti-establishment agitation.... The Society must beware of even appearing to be associated with those who are simply agitators or protectors. A reputation for obstructionism will inevitably prevent the Society's case from being considered."[46] Significantly, in 1971 the Standing Committee on National Parks was taken off the list of groups to receive consultation papers by the secretary of state for the environment on the grounds that it had become "too strident in its demands for reform of the National Parks Administration during discussions of local government reform."[47] It is of course possible for an interest group to pursue an "outsider" strategy, but this approach is risky and groups generally choose it only when they have no other options. In fact, the primary purpose of pursuing an outsider strategy is usually to secure insider status.

Civil servants have attempted to minimize the threat of environmental pressure groups to the existing patterns of clientelism by encouraging close cooperation between environmental organizations and government bodies that are peripheral to the major centers of decision making. Significantly, those bodies with whom environmental organizations tend to enjoy relatively close ties, such as the Countryside Commission, the Nature Conservancy Council, and the Historical Buildings Council, tend to have limited power and small budgets; their role is essentially advisory. Indeed, as Wyn Grant observes, one of the functions of the former semi-autonomous government agencies "is to create a kind of phony 'insider status' for some groups in order to reassure them that they have a point of access within the government machine."[48] On the other hand, those departments to which environmental organizations tend to find access more difficult, such as the Department of Industry, the Ministry of Agriculture, Fisheries, and Food, the Department of Trade, and the Department of Education and Science, tend to be extremely influential in shaping policies toward industry. These departments tend to be closely identified with the major sectional interest groups in British society, namely, industry, labor, and agriculture. The Department of the Environment appears to fall somewhere in between on both dimensions.[49]

## THE IMPACT OF ENVIRONMENTAL POLICY

While these arrangements have tended to mute both the intensity and the visibility of conflict over environmental policy in Britain, they have by no means eliminated it. In general, British environmental bodies have become much more involved in pressure-group politics over the last two decades and a number of policies have become the focus of considerable political controversy. The following four case studies include some of the more salient environmental controversies that have surfaced on the British political agenda over the last two decades and demonstrate the role of amenity groups in shaping their outcomes.

### The Third London Airport

In common with most other industrial countries, Britain has experienced considerable conflict in recent years over the efforts of public authorities to develop and expand their international airport facilities.[50] In Britain this conflict has been unusually prolonged. The government made its initial decision to develop a third international London airport in 1964; more than two decades later the government is still deciding where to put it. Each site proposed by the government has run into intense, well-organized, and highly sophisticated opposition on the part of local residents concerned about the negative impact of the construction and operation of a major airport near their homes or businesses.

The controversy had its origins in 1960, when an interdepartmental committee was formed to reappraise traffic projections for the London area. In 1964 the government published the committee's report: it recommended that a major international airport be constructed at Stansted. The committee's recommendation was accepted by the Labour minister of housing and local government, Richard Crossman, and an inspector was appointed to hold a public inquiry into "local objections relating to the suitability of Stansted for an airport and the effect of the proposed development on local interests." Following these hearings, the inspector submitted a report concluding that "it would be a major calamity for the neighbourhood if a major airport was placed at Stansted."[51] The government, however, decided to proceed despite the inspector's recommendations. The day the report was made public, the government issued a white paper stating that Stansted was indeed the most suitable location for a new international airport. This action created a public furor among both the Conservative and Labour benches in Parliament, and it was only by resorting to a three-line whip that the government was able to carry its motion.[52]

Given the intensity of parliamentary opposition—a day earlier more than 230 members had signed an early-day motion requesting that a royal commission be appointed to investigate the entire problem—the government found itself vulnerable to nonparliamentary pressures. These pressures rapidly materialized: a preservation association of local residents raised more than £25,000 from more than 13,000 individuals and launched an intense public campaign to force the government to reconsider its selection of Stansted. On February 22 Anthony Crosland, the newly appointed president of the Board of Trade, announced that the government would hold a new inquiry. A commission chaired by Justice (now Lord) Roskill was appointed "to inquire into the timing of the need for a four-runway airport to cater for the growth of traffic at existing airports serving the London area, to consider the various alternative sites and to recommend which site should be selected."[53] In March 1969 the Roskill Commission published a short list of four sites: Foulness, in Essex; Thurleigh, in Bedfordshire; Nuthampstead, in Hertfordshire; and Cublington (Wing), in Buckinghamshire.

A year and a half later the commission released its final report. It cost approximately £1.2 million and consisted of thirteen published volumes. Employing the most sophisticated cost-benefit techniques then available, the report recommended that the Cublington (Wing) site be chosen. As one of the commission members who dissented from its decision subsequently recalled, "all hell now broke out at Cublington." He added:

> There could be no question on this occasion of demanding another impartial inquiry. Even the most grudging critics of the Commission had to admit that it had been as thorough, open and impartial as could have been desired.... There was really no alternative now both for the resistance association and the local authorities to declare war, to announce their absolute determination not to have the airport, and to blazes with the Commission's calculations.[54]

The Wing Airport Resistance Association, which had been formed even before the commission had released its final report, organized an All Party Committee of Backbenchers to coordinate opposition to the selection of *any* inland site for a third international London airport. The committee succeeded in attracting the support of 219 MPs, drawn from all political parties.

With the cabinet divided and confronted with fierce hostility from Parliament, the government chose the path of least resistance. On April 26, 1971, the secretary of state for trade and industry announced that the government had decided to reject the committee's "two and one-half years of painstaking work"[55] and select Maplin Sands, off Foulness Island in the Thames Estuary, as the site of a third international airport. John Davies informed the Commons that "the irreversible damage that would be done to large tracts of countryside and to many settled communities by the creation of an airport at any of the three inland sites studied by the Commission was so great that it was worth paying the price involved in building the airport at Foulness." The prime minister claimed that by deciding to choose Foulness over Wing, "for the first time a government taking a major national decision has given pride of place to the protection of the environment."[56] Moreover, its decision to develop Maplin Sands represented one of "several bold projects on the Conservative agenda."[57] These projects included the construction of a tunnel connecting France and England, an expansion of the motorway system in and around London, and, most dramatic, the building of the Concorde.

This decision, however, did not resolve the issue. In 1974 the newly elected Labour government canceled the Maplin project, largely in response to the substantial unanticipated costs involved in land acquisition, construction, and provision of access to and from a seacoast site distant from London. A new interdepartmental study was commissioned to reassess future air traffic requirements for the London area. In 1979 Margaret Thatcher's government renewed plans to develop Stansted and a public inquiry on its proposal was held between 1981 and 1983. The inspector, Graham Eyre, recommended the construction of both a fifth terminal at Heathrow and the expansion of Stansted. The government subsequently issued a white paper that announced plans for a major expansion of airport facilities at Stansted.

What is striking about the pattern of opposition to the development of an airport on any inland site within reach of London is not simply its success over a period of nearly two decades but the kind and degree of political resources that opponents have been able to mobilize. In a sense, the nature of the British conflict over airport development represents, in a more capsulated form, the kind of political conflict over industrial and commercial development that has repeatedly taken place throughout the United States over the last decade: typically, well-educated members of relatively affluent communities have organized themselves to oppose changes in local land use that they regarded as threatening the value and current use of their property. Their efforts have been supported by national environmental organizations committed to protecting undeveloped land from industrial development.

The Wing Airport Resistance Association (WARA) represents virtually a textbook case of middle-class activism. WARA actually originated in Silverstone, which previously had been rumored to have been shortlisted by the Roskill Commission. In the autumn of 1968, three lawyers, two of whom had considerable professional experience in planning matters—one as an official of the National Farmers Union and the other as a member of the local executive of the County Landowners Association—formed an ad hoc committee to oppose the selection of Silverstone. A formal association was subsequently established, headed by two MPs, an alderman, and such prominent residents of the Silverstone area as Lady Hesketh, the chairman of the Northamptonshire County Council, the chairman of the National Provincial Bank, and Major General Vivian Street, who, in addition to having a distinguished military career, was a steward of the Jockey Club and well known for his charitable work with the Save the Children Fund. Thanks in part to the personal connections of its members, the Silverstone Airport Resistance Association—as it came to be called—was able to work closely with the local branches of the National Farmers Union, the Country Landowners Association, and the Council for the Protection of Rural England; it also succeeded in enlisting the cooperation of several county councils.

When the Roskill Commission published its short list of four sites in March 1969 and it was discovered that Silverstone was not included, the Silverstone association was not dissolved. Rather it became reconstituted, along with its constitution and bank balance, as the Wing Airport Resistance Association. (Wing is approximately sixteen miles southeast of Silverstone.) Within a month of Wing's appearance on the Roskill Commission's short list, a total of £25,000 had been collected or promised, WARA had retained a former Labour minister as its solicitor, and it had secured the services of an aviation and planning expert. The committee subsequently hired a noise expert and a public relations firm. It eventually raised a total of £62,000.

Sensitive to the charge that WARA represented little more than an upper-middle-class clique, reflecting the views of "the exclusive landed gentry and members of the hunting fraternity," the association launched a major effort to demonstrate grass-roots support for its position.[58] Thanks to local WARA committees that were organized in more than two hundred villages and towns, WARA was able to announce to the commission that a total of 61,766 local residents had formally joined their organization. It also endeavored to work with the existing network of voluntary organizations in the area. A total of forty-six local organizations declared their support for WARA. They included not only the local branches of the CPRE but groups as diverse as women's institutes, a disabled persons' club, gardening societies, and several local pigeon societies. The association's effectiveness was enhanced by the fact that several of its members had close personal or professional ties with prominent individuals in both Parliament and Whitehall. WARA's considerable financial resources, the political sophistication of its leadership, and the grass-roots support it was able to mobilize, coupled with the increased sensitivity of national politicians to environmental concerns, ultimately proved decisive in the government's rather abrupt reversal of the Roskill Commission's recommendation of Cublington on the grounds of cost and convenience.

## The Deposit of Poisonous Waste Act

In Britain as in other industrial societies, the media have played an important role in placing environmental issues on the political agenda. Their role was particularly decisive in securing the enactment of the Deposit of Poisonous Waste Act in 1972. The problem of the disposal of poisonous wastes first surfaced in Britain in 1963, when a large number of farm animals were killed by chemicals from a nearby factory. A committee established by the government shortly afterward recommended stricter controls, but its report was not issued until six years later, in 1970. The problem was subsequently studied by both the Royal Commission on Environmental Pollution and the Working Party on Refuse Disposal. While the Department of the Environment did conduct preliminary consultations with the Confederation of British Industry, no new regulations were either issued or proposed.

In November 1971, however, an event took place that finally put the problem on the political agenda. A lorry driver, after being rebuffed by local government officials, provided the Warwickshire branch of the Conservation Society with information about the unauthorized tipping of wastes in the Midlands. He produced samples of toxic materials, falsified delivery tickets describing toxic materials as innocuous, and evidence of conversations between drivers and the managers of a firm in which the latter promised bonuses for finding unsupervised sites. The local chapter forwarded a report to the secretary of state for the environment, identifying the companies involved and listing the sites at which drums of cyanide were believed to have been buried. It also called on the government to take legal action against the most prominent offender, to quarantine an area around a particular pit believed to hold 300 drums of cyanide, to organize reception depots for the temporary storage of toxic waste, and to launch an immediate public inquiry. The ministry, while conceding that existing legislation was inadequate, refused to make any substantive response to the society's requests. When approaches to local MPs also proved unproductive, a branch meeting voted to release some of the society's information to the mass media.

> The media reaction was suitably dramatic: most national newspapers gave the story priority coverage, and national and regional television . . . also provided considerable publicity. Fortuitously the story broke during the prelude to the United Nations Stockholm conference on the environment, when environmental matters were in vogue, and coincided with the publication of The Ecologist's "Blueprint for Survival" . . . . It was also the period when the Daily Mirror was pioneering the "Fight for a Cleaner Britain". . . . Interest in the issue was therefore temporarily maintained by both press and public and its news value fully exploited.[59]

The response of the government was immediate: the parliamentary undersecretary at the Department of the Environment, Eldon Griffiths, congratulated the society for its "public spirit and vigilance" and publicly reaffirmed the government's intention to review its regulations regarding refuse disposal procedures in order to prevent future abuses.[60] The National Association of Waste Disposal Contractors, which had been formed in 1968 precisely to impose some discipline on the industry's rather chaotic dumping procedures, expressed its concern about the dumping of poisons and indicated that it was working on a code of conduct that it planned to submit to the DOE for its approval. It also arranged a series of meetings with the department and urged the government to speed up its timetable for the introduction of new legislation.

Still no action was taken. Public attention soon shifted to a variety of other environmental "catastrophes" that were now being regularly highlighted by the media, and the pressure on the government to respond to this particular issue diminished. Even an extensive discussion of the issue in Parliament—one member contended that there was "plenty of evidence that cyanide and other toxic materials are being dumped in large quantities indiscriminately on tips all over Britain"—was insufficient to prod the government to advance its legislative timetable.[61]

For a second time the status quo was shattered by a scandal. In Nuneaton on February 24, 1972, thirty-six drums of cyanide waste were discovered on a derelict brickyard that was being used by local children as a playground. According to a local public health officer, this was sufficient cyanide to poison half the industrial Midlands. The next day the DOE's minister, Peter Walker, promised to "very quickly . . . make statutory the provisions which have already been agreed in the code of practice." The opposition spokesman, Dennis Howell, immediately offered Labour's support for emergency legislation, while Sir Bernard Braine, an MP who was connected with a leading waste disposal company, expressed the support of "responsible elements in the waste disposal industry" for "the earliest possible introduction of legislation with real teeth."[62] The Deposit of Poisonous Waste Bill was drafted and submitted to Parliament within two weeks, quickly approved by the Commons, subject to some fairly technical amendments in the Lords, and received the royal assent on March 30, 1972—only twenty-two days after the bill was originally introduced by the government. The government had not even taken the time to prepare a white paper or consult with the organizations affected by the bill. (In 1974 this law was superseded by the Control of Pollution Act.)

Public concern over this issue by no means disappeared with the enactment of this legislation. Three years later a survey revealed a total of fifty-one disposal sites that could be considered a risk to underground water supplies.[63] The British press has continued to carry reports of waste disposal sites that appear to threaten public health.[64] The emergence of this issue to a prominent place on the political agenda in the United States in the late 1970s and early 1980s also had an impact on public debate within Britain. In 1981 a lecturer in environmental health at Bristol Polytechnic suggested that "the balance of probability is that Great Britain will be very lucky if it can escape the legacy of its buried dumps." A British environmentalist added: "The question is not so much a case of 'Will a Love Canal happen here?' but 'When and how often?' " According to a government official, however, "we do not have the same problem as in the United States and it is invidious to make a comparison. Great Britain has a tradition of legislative control which just doesn't hold for the United States. We aren't smug about our record, but we are satisfied with it." While expressing some concern about the inadequacy of the disposal facilities available to many local waste disposal authorities, as well as the lack of any systematic monitoring of the production and disposal of hazardous waste in Great Britain, a 1981 select committee of the House of Lords indicated that it was "fairly confident" that Great Britain had no Love Canals festering in its soils. It cautioned, however, that "too many people are comforted by the belief that, because nothing much has gone wrong so far, nothing is likely to go wrong in the future. Constant vigilance will be needed or that comfortable belief could be rudely shattered."[65]

### Heavy Lorries

The political conflict in Britain over the government's policy in regard to heavy lorries is closely analogous to the dispute in the United States over federal standards for automobile emissions. In both cases the economic interests of an important industry were directly and successfully challenged by a political coalition spearheaded by environmental groups; indeed, legislation placing severe restrictions on automobile emissions was approved over the bitter opposition of the American automobile industry in the same year that the British government officially rejected the request of its domestic vehicle manufacturers that they be allowed to introduce lorries of more than 44 tons on Britain's roads. In neither case, however, did the decisions taken in 1970 resolve the issue; instead each government found itself subjected to continuous pressure on the part of vehicle manufacturers to reverse or modify its policy.

Between 1956 and 1964 the Ministry of Transport approved several increases in the permissible size and allowable weight of lorries operated on roads in Britain. In the mid-1960s vehicle manufacturers began to urge the government to approve another substantial increase in the gross operating weights of their units. In July 1969 the Society of Motor Manufacturers and Traders, after extensive discussions with various technical units in the Ministry of Transport, requested that the government revise its construction and use regulations to permit vehicles of 34, 44, and 56 gross tons. Their request was immediately supported by two other trade associations: the Road Haulage Association, representing the haulers, and the Freight Transport Association, which represented industrial and commercial organizations that operated their own transport.

All three trade associations contended that the proposed changes would increase the productivity of the hauling industry, thus benefiting themselves, their employees, and ultimately the consuming public. The manufacturers claimed also that these changes were essential if they were to produce vehicles for the export market, since gross operating weights of 38 tons or more were already standard throughout Europe. If the British home market failed to supply the volume base, the costs of developing and producing vehicles for sale overseas would be prohibitive.

They reminded the ministry that "successive Governments had urged the harmonization of international regulations affecting motor vehicle design," and that the modifications they requested would simply bring Britain in line with current general practices in Europe.[66] On the basis of their previous experience as well as their traditionally close relationship with officials in the Ministry of Transport, the associations had every confidence that their recommendations would be accepted by the government and approved by Parliament sometime before the end of 1969.

Public attitudes toward the heavy lorry, however, had become much more critical during the 1960s. This shift was due in part to the steady increase in both the number and size of such vehicles: those with gross operating weights of between 20 and 32 tons had increased from 34,000 in 1956 to 150,000 in 1968. This development, when coupled with substantial increases in the numbers of private cars, had led to severe strains on Britain's already inadequate road network. Many people held the large lorries "to be a source of danger, congestion, excessive noise and smoke and as the cause of significant physical damage to the urban and rural environment." These sentiments began to assume political significance following the appearance of a series of articles in the *Sunday Times*. Labeled "The Juggernauts," the pieces were highly critical of the road transport industry and articulated the case against the proposed increases on environmental grounds. They played a critical role in "effectively reinforc[ing] and then mobiliz[ing] the latent opposition to the heavy lorries which existed amongst its middle-class readership."[67] As a result, policies in regard to the weight and dimensions of lorries on British roads, which had formerly been settled through private negotiations between the Ministry of Transport and various trade associations, now became a subject of national political debate.

By the time Parliament recessed for the summer in 1969, other national newspapers had begun to give prominence to the issue; most editorial opinion was highly critical of the proposed increases. The National Council on Inland Transport released a study that contended that the proposed increases would triple the damage caused to road surfaces, while the chairman of the Noise Abatement Society accused lorries of having caused more damage to ancient buildings during the last fifty years than had been done in the previous ten centuries. The Pedestrians' Association for Road Safety also opposed the increases. The most important new participant in the debate, however, was the Faversham Society, one of the provincial amenity societies associated with the Civic Trust. Its secretary, Arthur Percival, approached the MP for Faversham and urged that the ministry consult the Civic Trust and the Council for the Protection of Rural England before any new changes in the lorry standards were adopted. The request was transmitted to the minister of transport, who agreed to give the two organizations consultation status on this issue.

The inclusion of amenity organizations on the list of groups to be consulted by the Department of Transport represented an extremely important political victory for British environmental interests; now the transport industry would no longer enjoy exclusive access to the civil servants responsible for formulating British transport policy. Shortly afterward the newly appointed minister of transport, Fred Mulley, told the press that he would not be able to make an early decision on the weights issue because he would now require additional time to consult the new organizations that had become participants in the decision-making process.

The CPRE, which had long been concerned about the damages to the countryside caused by additional motorway construction and the increased use of heavy lorries, responded to the minister's statement by opening a substantial press campaign. It began with a public letter from its chairman, Lord Molson, endorsing the position of the Civic Trust. In a subsequent statement to the press, the committee dramatically predicted that "if lorry loads were to be further increased, then life would become intolerable."[68] The political position of the amenity societies was significantly strengthened when a number of local governmental authorities responsible for road maintenance expressed their concern over the escalating road repair bills caused by increased heavy lorry traffic. In addition, the Royal Institute of British Architects urged that the industry's proposal be subjected to thorough cost-benefit analysis.

Both sets of arguments—environmental and economic—were voiced in both houses of Parliament in December 1969. Subject to widespread constituent pressures, MPs of both parties informed Fred Mulley that there would be "widespread dismay . . . if some of the proposals being suggested by the road haulers [were] accepted." The campaign against approval of the increased weight allowance reached its climax in September 1970 with the presentation of an exhaustively documented memorandum compiled by the Civic Trust and presented to the Ministry of Transport. Based on the submissions of 300 local civic societies—nearly 45 percent of the total number of provincial societies registered with the Trust—the document "contained the most comprehensive catalogue of lorry nuisance and damage that had hitherto been compiled and provided the Ministry with a ready answer to the 'cogent economic argument' of the transport industry."[69] It listed fifty-odd categories of nuisances caused by lorries; the image the report presented was of bulls in the china shops of English towns and villages.

By the time the ministry was due to make its decision, "it had become clear that a decision favoring the road-transport industry would fly in the face of a substantial segment of public opinion." Moreover, in the eyes of the environmental movement, the issue had become the acid test of the intentions of the newly elected Conservative government toward conservation and environmental regulation. After a month of intensive press speculation and numerous conflicting rumors, on December 16, 1977, the minister of transport informed Parliament of his decision not to allow any increase in the maximum weights of goods vehicles. Indicating that he "shared the concern about the effects of heavy lorries on the environment which had been expressed to him from many quarters," he proposed a series of new regulations designed to reduce the exhaust level and noise of lorries.[70] The only concession made to the transport industry was a promise to draft regulations permitting marginal weight increases for certain vehicles within the 32-ton limit. While graciously conceding defeat, the leaders of the industry were extremely disappointed. Their informal agreement with the Ministry of Transport had been overridden by the newly established minister for the environment, eager to establish his credentials with a politically potent environmentalist constituency.

This decision did not resolve the issue, however, but rather set the stage for further controversy. The first threat to the British government's 1970 decision was occasioned by Britain's negotiations to enter the European Economic Community (EEC). The British negotiators immediately found themselves under pressure to adhere to the clause in the Treaty of Rome calling for a common transport policy. Since the Common Market norm stood at 38 tons, any reasonable compromise presented by the EEC Commission to the Council of Ministers would have required Great Britain to increase its allowable tonnage immediately, thus effectively annulling the government's decision less than a year after it had been made. Following a complex series of negotiations, however, the British representative at the Council of Ministers, faced with overwhelming pressure from both the opposition in Parliament and the government's own backbenchers, insisted that Britain be allowed to maintain its restrictive standards. Not wanting to antagonize the British public in the midst of the delicate negotiations then pending surrounding Britain's membership, the Six yielded and British policy was able to stand with only minor modifications.

Following the successful British resistance to the EEC Commission's efforts to harmonize vehicle weights, amenity groups in Britain began a campaign to reduce the nuisance caused by lorries already permitted on British roads. Specifically, the Civil Trust encouraged the organizations registered with it to pressure the government to introduce a national system of mandatory designated routes that would keep heavy commercial vehicles away from town centers, traffic areas, residential neighborhoods, and country villages. Citing the complex administrative and political problems involved in fully implementing this proposal, the government nevertheless partially responded to the Trust's requests by issuing a circular advising local authorities to install a series of signs to assist lorry drivers in finding suitable roads. The Trust welcomed this concession but characterized it as "piecemeal and makeshift" and continued to press for a comprehensive system of mandatory routes.[71]

The transport industry accepted in principle the need for designated routes—though favoring as flexible a system as possible—on the grounds that its willingness to accept restrictions on the places where lorries could be used would increase the likelihood that both the government and the amenity interests would support higher weight limits. The industry launched a major public relations campaign designed to inform the public of its contribution to the national economy. The Road Haulage Association and the Freight Transport Association cooperated in publishing a report written by an academic economist titled *Living with the Lorry*, while the FTA published a pamphlet titled *Lorries and the Environment*, which argued that "in terms of the environment, the size and weight of lorries is much less significant than where they go."[72] Another report, *Lorries and the World We Live In*, was produced by the road haulage industry at the behest of the Ministry of Transport; it candidly acknowledged the environmental damage for which lorries were responsible but concluded that most such problems could be resolved by a combination of adequate roads and sensible traffic management. Accordingly, the industry continued to press for increased government expenditures on roadways.

A decade later, the issue of the appropriate weights of lorries once again emerged on the political agenda. A committee chaired by Sir Arthur Armitage recommended that lorries of up to 42 tons be permitted on British roads. Faced with intense parliamentary opposition, the Thatcher government compromised and announced a 38-ton limit, thus finally bringing British standards in line with those of the other member states of the European Community.

The conflict over the maximum size of lorries both reflected and provoked an important shift in public debate about national transportation policy in general. In contrast to much of the rest of the European Community, British commerce remains unusually dependent on road transportation. The percentage of goods transported by road increased steadily during the 1960s; by 1971, 85 percent of all goods shipments were made by lorries. In America the automobile had become a popular symbol of the nation's "distorted and wasteful" transportation policies; the lorry came to occupy an analogous place in the world view of the British environmental movement. In both countries the intense conflict over public policy in regard to specific regulations for the problem vehicles in the late 1960s and early 1970s became transformed into a more general political and ideological dispute over the nature of transportation policy.

In the early 1970s both CoEnCo and the CPRE called for a major review of national transport policy. A CPRE conference called "Transport—Coordination of Chaos" approved a resolution calling on the government to reexamine current transport investment priorities and attempt to calculate the "acknowledged disamenities of roads and road traffic." The Civic Trust also urged a major review of British transport policy, arguing that "concentrating on road transport may in the long run prove as much a disaster for the economy as [for] the environment."[73]

In late 1971, conservation forces began to coordinate their efforts to counterbalance the political influence of the roads lobby. A Transport Reform Group (TRG) was formed from various affiliates and associates of the CPRE and the Civic Trust in order to assist local groups opposing expansion of the British motorway system and to pressure the Department of the Environment to allow organizations opposed to motorways greater access to the decision-making process. Paralleling the TRG's formation and partially inspired by it, a number of local organizations involved in opposing various road schemes began to form regionally based alliances in order to exchange ideas and strategies and increase their political visibility and influence. In 1972 a Transport and Environment Group, consisting of representatives of more than twenty-five environment- and amenity-conscious organizations—the latter included such groups as the National Council of Women and the National Union of Students—was formed to persuade the political parties to reevaluate their transport policies and give increased priority to public rather than private forms of transportation.

While the Transport and Environment Group formally represented more than 5 million people, its political resources were limited. Its political effectiveness, however, was substantially enhanced by the alliance it developed with the railway unions and the Railway Industry Association. Organized in 1973 at the initiative of the National Union of Railwaymen (NUR), Transport 2000 consisted of a loose federation of twenty major conservation groups, unions, and railway industry organizations. Primarily financed and staffed by the NUR, it established twenty regional groups, each corresponding to a local regional authority. Working through the Transport Industries Committee of the Trades Union Congress, it urged the Department of the Environment to improve the quality and quantity of urban transport services. The Labour party, then in opposition, indicated its support for the general aims of the coalition, and a party "green paper," published in the summer of 1973, called on the government to reverse the trend of reduced public support for railways on environmental grounds. The Conservative government responded with a white paper that addressed the need to improve urban transport but refused to undertake a comprehensive reexamination of current road-building projects on the grounds that further delays in motorway construction would be counterproductive.

Nonetheless, the antimotorway forces in British politics have enjoyed considerable success at both the local and national levels. A number of motorway projects have been halted as a response to vigorous community opposition. In 1973 the Conservative government announced that it planned to transfer £200 million from the road program to the railway system over the next five years; expenditures on railways were slated for a 27 percent increase. The government also decided to introduce a new and simplified system of transport grants to local governments in order to enable local authorities to switch funds from road construction to public transit.

## LEAD IN PETROL

Since the passage of the Clean Air Act (1956), health issues have not figured prominently in the debate over British environmental policy. There is, however, one notable exception: the controversy over the lead content of petrol. The Control of Pollution Act of 1974 gave the Department of Transport the power to regulate the content of fuels used in motor vehicles. Faced with a steady accumulation of evidence that high blood concentrations of lead presented a health hazard, particularly to children, the British government began to reduce the level of lead permitted in petrol. In 1972 it was reduced from 0.84 grams per liter to 0.64 grams, and in 1974 to 0.55. A few years later it was further reduced to 0.45 grams per liter, and in 1980 the government announced that this figure would be reduced to 0.4 by 1981, thus bringing Britain into harmony with the EEC standard. The first two reductions were agreed to voluntarily by the petroleum industry; the 1980 cut was effected by the issuance of regulations.

These reductions, however, failed to satisfy those who insisted that the only responsible policy was for Great Britain to ban the use of lead in petrol altogether, thus bringing British policy more in line with those of the United States, which had required all new cars to run on lead-free gasoline since 1973, and Japan, which also had banned the use of lead in new cars during the 1970s. In 1981 a public campaign was launched to pressure the British government to require that all new cars sold in Great Britain run on leadfree petrol. Its supporters, who included a large number of national environmental, public health, and social welfare organizations, contended that there was overwhelming scientific evidence that lead posed a threat to the health of children. Their campaign was supported by the British Medical Association and endorsed by editorials in national newspapers. The *Times* argued:

> We should not have to wait until the very last mathematical correlation has been established to announce proudly that there is final proof that children have continued to be blighted while the research was concluded. The balance of risk is clearly such as to justify the maximum control on the emissions of lead poisons.[74]

Within a year, CLEAR—the Campaign for Lead-Free Air—had been endorsed by both the Social Democratic and Labour parties, as well as by the Trades Union Congress, more than 100 local authorities, and 200 members of Parliament. Survey data indicated that the majority of British doctors and 90 percent of the British public supported the ban.

The government initially refused to go beyond limiting the lead content of petrol to 0.15 grams per liter by 1986—a decision it announced in May 1981. This policy was based on a study conducted by a working party of Great Britain's Department of Health, which concluded that there was no convincing clinical evidence that blood concentration levels of lead below 35 to 40 micrograms per milliliter had any negative effect on the health of children. It noted that while 5 of 231 preschool children tested did have concentration levels above 35 micrograms, these elevated levels seemed to result from exposure to sources of lead other than automobiles, such as water from lead pipes. Their position was supported by an article in the *British Medical Journal*, which asserted: "There is, so far as we are aware, no new evidence to justify [the argument] that there is a strong likelihood that lead in petrol is permanently reducing the IQ of many of our children."[75] *The Economist*, while suggesting that perhaps "Britain should play it safe and go for a ban on lead despite the cost and despite the scientific uncertainty," nevertheless concluded that "it is a disservice to informed debate...to pretend that the medical uncertainties have now been banished. Or to pretend that lead-free petrol is necessarily the best health buy for 200 million pounds."[76]

In 1983 the Royal Commission on Environmental Pollution released its own recommendation. While noting that the average concentration of lead in the blood of the British population was roughly a quarter of that needed to produce overt symptoms of lead poisoning, it nonetheless concluded that there was no justification for the setting of arbitrary figures for safe concentration levels. It recommended that, beginning in 1990, all cars produced in Britain be required to run on leadfree petrol. The cost of this change to the average motorist was estimated to be approximately £10 a year.

Immediately following this report, the Thatcher government reversed its position and agreed to phase out the use of lead within the next decade. While opposed by the British automobile industry—which will, however, have six years to adjust to it—the government's decision was supported by the Petroleum Industry Association. Underlying the latter's stance was its judgment that it was cheaper to eliminate lead altogether than to reduce it to an extremely low level. However, Britain's decision cannot go into effect until the European Community changes its rules, which currently do not permit any member country to require the level of lead to fall below 0.15 grams per liter.

## CONCLUSION

In Britain as in all industrial societies, environmental regulation has come to occupy an increasingly important place on the domestic political agenda over the last two decades. Britain already had an extensive network of environmental organizations before the mid-1960s; during the 1970s they became more politicized and their influence on public policy increased. The four case studies presented in this chapter illustrate both the saliency of environmental issues within the context of British politics and the relative influence of British amenity organizations on public policy. Yet their significance needs to be kept in perspective. In fact, relatively few environmental policies in Britain have come to the attention of either Parliament or a governmental minister; most are decided in more private and less visible forums, usually through negotiations between civil servants and representatives of industry from which amenity groups and the public are often excluded. The British government's relationship with environmental organizations, like that with industry, has remained largely a cooperative one. For the most part such organizations continue to work closely with Britain's permanent civil servants on both the formation and implementation of policy. For all the changes in British environmental policy described here, the pace of policy innovation remains modest, and high priority is still accorded the values of compromise and consensus. While these cases reveal some of the ways in which amenity values and interests have affected British regulatory policy, the extent of their influence remains limited. Relatively few issues have been resolved in a manner so favorable to British amenity interests; compared to the major sectional interests of British society—industry, labor, and agriculture—the influence of environmental constituencies remains marginal. And these constituencies continue to enjoy far less access to the policy process. Each of these points will emerge more clearly in the following two chapters.

CHAPTER TWO

# The Politics and Administration
# of Pollution Control

The British government's distinctive approach to environmental regulation emerges most sharply in the area of pollution control. While in Britain, as in other industrialized democracies, policy in this area is in a state of flux, it is possible to identify several distinctive elements: an absence of statutory standards, minimal use of prosecution, a flexible enforcement strategy, considerable administrative discretion, decentralized implementation, close cooperation between regulators and the regulated, and restrictions on the ability of nonindustry constituencies to participate in the regulatory process. As we shall see, each of these characteristics stands in sharp contrast to the strategy adopted by the American government to control industrial emissions.

## The Legal and Administrative Framework

The responsibility for controlling stationary sources of air pollution is divided between central and local authorities. The more important and complex stationary pollution sources are regulated by Her Majesty's Industrial Air Pollution (Alkali) Inspectorate.[1] As recently as twenty-five years ago, the Inspectorate played a relatively minor role in controlling air pollution in Great Britain: in 1957 it had only nine inspectors. In 1958, following the recommendations of the Beaver Committee, the number of plants under its jurisdiction more than doubled. It was made responsible for controlling the emissions from power plants and from iron, steel, copper, aluminum, gas, coke, and ceramic works. Between 1970 and 1979 the number of "registered" works was further increased

70

about 25 percent. As of 1979, the Inspectorate was responsible for controlling the emissions produced by approximately 2,000 "registered" works throughout England, Wales, and Northern Ireland. (A similar though separate body, Her Majesty's Industrial Pollution Inspectorate, has jurisdiction in Scotland.) These plants represent approximately 3,000 industrial processes that have given "rise to particularly noxious or offensive emissions or are technically difficult to control";[2] collectively they consume approximately 75 percent of the fuel burned by industry in the United Kingdom.

In 1975 the Inspectorate was placed under the jurisdiction of the Health and Safety Commission, a quasi-autonomous body that is also responsible for health and safety at the workplace. The Inspectorate continues, however, to work closely with the Department of the Environment. Its basic statutory authority remains the Alkali, Etc. Works Regulation Act of 1906, which provided that no plant may operate a scheduled process without first obtaining a certificate of registration, which it must renew annually. It also required that upon registration all new plants must adopt any pollution-control measures deemed necessary by the chief inspector. The Inspectorate currently has about fifty inspectors, based either in London or in fifteen area offices in England and Wales. Each inspector makes an average of seven visits per year to each registered plant under his supervision. All hold degrees in either chemistry or chemical engineering and have had at least five years of experience working in industry.

Domestic sources of air pollution as well as pollution from the 30,000 to 50,000 plants not registered by the Inspectorate are under the jurisdiction of some 450 environmental health departments. These departments, which are appointed by various local governmental authorities, have a wide range of responsibilities, including food inspection, pest control, and the maintenance of drainage systems; pollution control accounts for about 10 percent of their activities. Part II of the Public Health Act of 1936 granted district councils and London boroughs the power to inspect their jurisdictions for "any dust or effluvia caused by any trade, business, manufacture, or process" that was "prejudicial to the health or a nuisance to the inhabitants of their district," and to take action to abate it.[3] Parliament has subsequently made the responsibilities of local authorities for controlling air pollution more specific. The Clean Air Acts of 1956 and 1968 defined four specific offenses for which local authorities may prosecute: the breach of a smoke-control order, the emission of dark smoke, the emission of grit and dust, and the breach of standards regarding new boilers and chimneys. They also have the right to approve chimney heights and in some cases to approve abatement equipment before its installation. In practice, much of their attention has been devoted to controlling the dark smoke produced by the burning of coal, particularly in households.

The effort to control water pollution in Great Britain dates from 1876, when Parliament enacted the Rivers Pollution Prevention Act. This law, however, was not enforced by local authorities. In 1947, after noting that "the scandal of river pollution is fully recognized and must be dealt with," the Attlee government introduced a bill consolidating 45 fishery boards, 53 catchment boards, and 1,600 local pollution authorities into 32 river boards in England and Wales.[4] The boards were given the right to prescribe standards for the rivers under their jurisdiction. In addition, all new or altered discharges were now required to receive specific consent. In 1960 their jurisdiction was extended to tidal waters, and in 1961, consents were required for all existing discharges into rivers. The 1963 Water Resources Act subsequently established 29 river authorities, which were also made responsible for water conservation.

These reforms, however, proved unsatisfactory. An official report published in 1970 revealed that nearly 60 percent of sewerage works were, on a yearly average basis, operating outside the consent conditions of the river authorities, while more than 3,000 plants were producing effluent "inferior to what could reasonably be expected by the use of modern treatment methods." The reason for the lack of enforcement was a simple one: the majority of the members of the river authorities were appointed by county borough and district councils. Since they were required to authorize expenditures for additional sewage treatment facilities, they were hardly likely to impose additional financial burdens on themselves. Besides, as the Working Party on Sewage Disposal pointed out, "it is difficult for river authorities to insist on the standards required for industrial discharge if the discharges of the local sewage authorities are not up to standard."[5]

By the late 1960s, the problem of water quality in Britain had become critical. Not only was the public becoming increasingly concerned about the poor quality of British rivers, but community opposition to the construction of new reservoirs exacerbated the difficulty of increasing the water supply by this means.[6] Britain faced an impending water shortage that could no longer be resolved simply by increasing its supply; instead priority now had to be placed on finding a way of reusing "dirty" water. As a result, improving water quality became an economic as well as an environmental concern.

In 1970 the newly established Department of the Environment proposed a radically new two-pronged approach to the administration of water resources in Britain. First, water would be removed from the control of local governments and placed in the hands of a group of regional authorities. Second, water supply, sewerage and sewage disposal, and water conservation would become the responsibility of a single multifunctional organization. The DOE's proposals were subsequently modified following negotiations with local authorities, who were concerned about the loss of their power, and with agricultural interests, who worried that land drainage would be accorded a lower priority under the new arrangement.[7]

The legislation submitted to Parliament adopted a managerial approach to the control of water, with emphasis on efficiency and rationality. It consolidated the 198 separate statutory companies responsible for water supply, the more than 1,300 county borough and district councils in charge of sewerage and sewage disposal, and the 29 river authorities responsible for water conservation into ten regional water authorities (RWAs)—nine in England and one in Wales—each corresponding roughly to a separate water drainage system. They were given responsibility for all aspects of water management, including conservation, water supply, sewerage and sewage disposal, land drainage, fisheries, water recreation, flood prevention, and the control of emissions into rivers. The Control of Pollution Act of 1974 subsequently granted them the power to control discharges into estuaries and the sea as well as into rivers.

The Water Act (1973) provided for a majority of the members of each RWA to be appointed by the county and district councils within its area, with each represented in proportion to its population. The remaining members were to be appointed by either the secretary of state for the environment or the minister of agriculture, fisheries, and food. Each RWA was, in principle, meant to be financially self-supporting, receiving the bulk of its revenue in the form of charges for such services as sewage disposal and water supply. The 1973 legislation also established an advisory National Water Council to coordinate the operations of water authorities and function as a liaison between them and the Department of the Environment. Its membership was comprised of the chief executives of the RWAs and various officials appointed by the secretary for the environment.

This structure was subsequently altered by the Water Act (1983), which abolished the National Water Council and established in its stead a Water Authorities Association. This administrative change, however, had nothing to do with environmental policy. It was initiated by the Thatcher government to reduce the bargaining power of the national union representing water workers. To improve the efficiency of the system, the membership of each RWA was reduced by about two-thirds and ministers were given exclusive authority to select them. At the same time, the government established a consumer advisory council for each RWA.

In addition to air and water pollution, the Department of the Environment regulates noise pollution—other than that from aircraft—the spilling of oil and chemicals on beaches, and the disposal of solid and radioactive wastes. The remaining pollution-control responsibilities of the British government are divided among other ministries. All pollution from civil aviation activities as well as marine pollution from oil are under the jurisdiction of the Department of Trade, while the Ministry of Agriculture, Fisheries, and Food supervises the use of pesticides and dumping at sea. The Health and Safety Commission is responsible for environmental problems related to occupational health and safety. In 1971 a Central Scientific Unit on Environmental Pollution was established within the cabinet office and subsequently transferred to the Department of the Environment. (Its name has since been changed to the Central Directorate of Environmental Protection and its purview has been slightly restructured.) It has no direct regulatory responsibilities, but attempts to coordinate all the various pollution-control efforts of the British government. It regularly makes appraisals of pollution problems and provides the management capacity for the overall monitoring and assessment of the quality of the British environment.

The secretary of state for the environment functions as the final legal arbiter for the pollution-control efforts of the British government. Decisions of both the Alkali Inspectorate and the regional water authorities may be appealed to him. In such cases, he may at his discretion choose to hold a formal hearing, but he is not bound to follow the recommendations of the hearing officer. In unusual cases, the decisions of these bodies may be referred to the High Court for adjudication, and the minister is bound to follow its decision. But his consent is required before the High Court can hear the issue in the first place. Moreover, his permission is required before local authorities can bring a nuisance action against companies that have produced "excessive pollution." It is possible for individuals to bring suits against polluters under the common law doctrine of nuisance, and more than one thousand such suits have in fact been brought in recent years, mostly by local angling associations. While many of these suits have been successful, they have affected water quality only in those rivers and streams that are privately owned.

In general, Parliament has left both the formulation and the implementation of pollution-control policies to the Department of the Environment. British pollution-control statutes tend to confine themselves to establishing board policy objectives, and to the extent that they have involved important changes in policy, negotiations among interest groups primarily take place before the submission of legislation. Moreover, the secretary of state for the environment enjoys considerable discretion in deciding when, how, and even if specific statutory provisions are to be enforced. A decade after the passage of the Control of Pollution Act (1974), for example, several of its key provisions in regard to disclosure had not been implemented.

Over the last three decades, however, Parliament has periodically played a role in shaping British pollution policy: there was intense debate over the enactment of the Clean Air Act (1956), a parliamentary committee conducted an investigation following the breakup of the *Torrey Canyon*, and parliamentary pressure helped force the government to expedite consideration of the Deposit of Poisonous Wastes Bill.[8] Amenity groups were able to use their contacts in Parliament to offer amendments that strengthened both the Water Act of 1973 and the Control of Pollution Act of 1974. In the early 1980s the Commons passed a resolution urging the government to restrict the lead content of petrol, and parliamentary committees have from time to time issued reports critical of the regulatory policies of the Alkali Inspectorate. In addition, the House of Lords has been the setting for extremely sophisticated discussions of the impact of Britain's membership in the EEC on its pollution-control policies.

Within the government itself, except when new legislation is being considered, the role of the cabinet and the secretaries of state tends to involve primarily determining the funding levels of the Alkali Inspectorate and indirectly, through their control over local rates (property taxes) and the borrowing authority of nationalized industry, those of the regional water authorities. (The RWAs are considered nationalized firms.) The secretary of state for the environment also appoints the members of each regional water authority. In addition, ministers have periodically convened committees of inquiry to address specific pollution problems. On the whole, however, policy implementation at the national level is left almost entirely to civil servants and direct ministerial involvement is unusual. While environmental health departments are accountable to the elected officials of local governmental units, the Alkali Inspectorate has historically operated with almost complete autonomy, and its transfer from the Department of the Environment to the Health and Safety Commission has not altered its freedom. All its officials, including the chief inspector, are civil servants.

## The Role of Standards

What sharply distinguishes the British government's approach to pollution control not only from that of the United States but from that of any continental nation is its flexibility. The British make less use of legally enforceable environmental quality or emission standards than does any other industrial society. They attempt to tailor pollution-control requirements to meet the particular circumstances of each individual polluter and the surrounding environment. The secretary of state for the environment wrote in 1977:

> Except in clearly defined cases, we believe it is better to maintain gradual progress in improving the environment in light of local circumstances and needs than to operate through the formation of rigid national emission standards which may be in particular circumstances either unnecessarily harsh or insufficiently restrictive.[9]

An official department publication adds: "[Our] pragmatic approach permits the establishment of individual standards for polluting emissions from particular factories which can be made continually more stringent in the light of technical advance and of changing environmental needs, but allows greater flexibility than statutory standards."[10] Lord Ashby, a distinguished biologist and the first chairman of the Royal Commission on Environmental Pollution, argues that by not laying down rigid standards, the British government "gives the authorities responsible for administrating the legislation discretion to adjust what are really permits to pollute according to the circumstances of the place and time and industry or corporation concerned."[11] By contrast, as the director of the Environment and Consumer Protection Service of the European Commission noted, "the 'Continentals' tend to believe more in standards defined on the basis of best technical means and applied through mandatory instruments."[12]

Underlying these differences is a distinctive British conception of the appropriate purpose of legal requirements: the British are reluctant to adopt rules and regulations with which they cannot guarantee compliance. As a British document on the impact of Britain's membership in the European Community notes:

> British environmental legislation . . . has a tendency to caution, with only small steps being taken at a time, and with care being taken that demands are not made that cannot be realized. . . . In other countries it is much more common for legislation to be used to force the pace of change, sometimes making demands that cannot be immediately achieved, but with timetables for doing so and with pollution levels centrally specified.[13]

Lord Ashby makes a similar point in contrasting British and American environmental policy.

There is a striking contrast between the style of environmental legislation in Britain and the USA. British lawmaking is piecemeal, reluctant, never wanting to pass statutes which promise more than could be performed. Some American lawmaking is unlike this; for example the U.S. Water Pollution Control Amendment Act of 1972, which aspired to make all rivers fishable and swimmable by 1985.[14]

Michael Hill, who has written extensively on British air-pollution regulation, adds: "The British system . . . minimizes that disrespect for administrative action which arises where political 'goals' and practical implementation achievements are markedly out of line."[15]

The 1906 Alkali Act did establish specific emission standards for hydrochloric acid and sulfuric acid. They still remain in force, but subsequent legislation has omitted any reference to them. The Inspectorate has, however, established "presumptive limits" for many of the pollutants emitted by the industries under its jurisdiction. If a firm has complied with these limits, it is presumed to be exercising appropriate control over its emissions. Although a company's failure to comply with the limits may be used as evidence in various legal proceedings, the limits are not legally binding. The Inspectorate may require new plants to keep their emissions below its published "presumptive limits" while older facilities are generally given additional time to comply with them, depending on the life expectancy of their existing pollution-control equipment and of the plant itself. At the local level, the Clean Air Act of 1968 does include a reasonably specific emission standard related to dark smoke: it is defined as shade 2 or above on the Ringleman chart. The specificity of this statutory language is highly unusual, however, and local authorities exercise considerable discretion in determining how strictly to enforce it. A few local authorities have been given permission to regulate the sulfur content of fuel burned within their jurisdictions, but the Department of the Environment has specifically forbidden them to establish legally binding air-quality standards.

While the Alkali Inspectorate does take into consideration ground-level concentrations in determining emission requirements, British pollution-control authorities have strongly opposed the establishment of uniform ambient air quality standards, a common practice in Europe, the United States, and Japan. The Fifth Report of the Royal Commission on Environmental Pollution argued that "ambient standards were generally impracticable because of the difficulty of relating them to specific sources," and remarked on "the difficulties even in establishing 'bands' of desirable air quality and in monitoring air quality sensibly."[16] The British also make extremely limited use of effluent standards to regulate water quality; instead the government relies extensively on its control over industrial siting as a means of affecting the quality of each particular body of water. Britain also has no uniform national standard for drinking water. The government requires merely that potable water be "wholesome"; it is up to each regional water authority to decide what constitutes wholesomeness.

The Water Act (1973) continued the system of consents that had been established more than two decades earlier. All sources of effluents discharged into virtually all bodies of water are first required to be granted a consent by a regional water authority. A consent is individually negotiated with each polluter to reflect the nature of the substances being discharged, the capacity of the water system to absorb them, and the ultimate disposition of the water supply. It may either be unconditional or subject to such factors as the duration, rate, quality, and composition of the discharge. While on the whole consents tend not to vary greatly, even among very different effluents, in practice the way in which they are enforced depends heavily on the capability of the receiving environment. According to one senior official, "treating discharges alike in terms of . . . the quality of the discharge [i.e., an emission standard], as distinct from the effect on the environment, is absolute nonsense."[17] At the same time, inspectors do consider themselves under a moral obligation to treat similar discharges in a roughly similar fashion so that no manufacturer is placed at a competitive advantage or disadvantage. In any event, the RWAs do not attempt to enforce consents strictly. In effect, like the presumptive limits employed by the Alkali Inspectorate, consents function less as a standard than as a basis for further negotiation and bargaining. Thus "the river authority combines the functions of both legislative and enforcement agency."[18] Seventy percent of industrial discharges flow into sewerage plants controlled by the water authorities. While Britain does not have a formal system of effluent charges, companies do pay for the use of these facilities in proportion to the volume and strength of their effluents; thus they have a financial incentive to reduce their emissions.

In 1976 the National Water Council recommended that Britain adopt a modified system of water quality objectives (WQOs). Under this system, which is now being put into effect, each RWA is responsible for adopting a quality objective for each body of water under its jurisdiction. Each objective is to be based on the specific purposes for which that body of water is intended. Accordingly, water intended to be recycled for domestic consumption will be required to meet stricter standards than downstream river water that will flow into the North Sea without further use. Consent conditions are being modified to reflect those WQOs. Thus

Britain is moving toward a system of water quality objectives, but they will not be uniform and will serve primarily as guidelines for the issuing of consents. This important shift in policy, which in other countries would have required legislation, was formulated and implemented entirely by the National Water Council.

## BEST PRACTICABLE MEANS

The clearest expression of the British approach to pollution control is the concept of "best practicable means" (bpm). The principle underlying this term was first enunciated in the Alkali Act of 1874, which required owners of registered works to install that technology of abatement which was "reasonably practicable and technically possible to prevent the emission of noxious gases and render those discharges harmless."[19] It was explicitly reaffirmed in the Health and Safety at Work, etc. Act of 1974, which placed owners of registered works under a duty to use "the best practicable means for preventing the emission into the atmosphere from the premises of noxious or offensive substances as may be so emitted."[20] Bpm essentially represents a highly complex balancing act. As the chief alkali inspector noted in 1957,

> there must be compromise between (1) the natural desire of the public to enjoy pure air, (2) the legitimate desire of manufacturers to meet competition by producing their goods cheaply and therefore to avoid unremunerative expenses, (3) overriding national interests. The answer to these opposing interests lies in the honest use . . . of the best practicable means . . . .[21]

While bpm legally refers only to the standards imposed by the Alkali Inspectorate, it underlies virtually the entire British pollution-control effort. According to an official statement from the Department of the Environment, "authorities, both central and local, are expected to operate on the philosophy that standards should be reasonably practicable."[22] The term "practicable" has never been clearly defined; indeed, the Alkali Inspectorate is itself considered the sole judge as to whether the plants and processes under its jurisdiction are employing the best practicable means of controlling their emissions. In practice, however, it has come to encompass local conditions and circumstances, the state of technological knowledge, and the costs of pollution control. "Means" refers not only to the installation of pollution-control equipment itself but also to its maintenance and design, as well as the operation of the plants themselves. The *Notes on Best Practicable Means* that the Inspectorate publishes for each industry or process under its jurisdiction are extremely detailed and rather technical. They describe how emissions are to be measured, the scientific basis for the establishment of presumptive limits, the kind of abatement equipment to be employed, and even the appropriate means of maintaining it. (In this sense, they are akin to the New Source Performance standards adopted by the Environmental Protection Agency.)

Except when there is a clear and imminent danger to public health, British authorities have generally been reluctant to impose greater costs of pollution control than the plants under their jurisdiction can reasonably afford. The chief alkali inspector stated in 1970:

> If money were unlimited there would be few problems of air pollution control that could not be solved fairly quickly. We have the technical knowledge to absorb gases, etc. The only reason why we still permit the escape of these pollutants is because economics are an important part of the word "practicable." Most of our problems are chequebook rather than technical.[23]

According to Keith Hawkins' study of the enforcement activities of the RWAs, "economic incapacity is recognized by field inspectors as a 'genuine' reason for non-compliance."[24] Accordingly, one reason that formal cost-benefit analysis has played a small role in British pollution-control policy is that the costs of compliance are already built into both the rulemaking and enforcement processes. In practice, the severity of enforcement varies not only from region to region, depending on economic circumstances, but also over time. Thus enforcement has recently become less vigorous as a result of the current hardships experienced by many sectors of British manufacturing. No change has been made in the regulations themselves, merely in the deadline for compliance with them. The Inspectorate declares: "The expression 'best practicable means' takes into account economics in all its financial implications, and we interpret this not just in the narrow sense of the works dipping into its own pockets, but including the wider effects on the community."[25]

The other critical component of bpm involves the state of technological knowledge. British officials tailor their pollution-control requirements to the technical capacity of industry to reduce or disperse emissions; requirements are not imposed until officials are satisfied not only that industry has the technical ability to meet them but that the costs of introducing this technology are reasonable. As one regulatory official put it, "the technically possible would be impracticable if the costs were so high that the manufacturing operation were thereby rendered unprofitable or nearly so."[26] On the other hand, as pollution-control technology progressively improves, emissions requirements are gradually tightened. The presumptive limits established for each industrial process are reappraised periodically, generally once every ten to fifteen years. Accordingly, bpm, in the words of a former chief alkali inspector, should be regarded as "an elastic band" that can be tightened as "science [develops] and [places] greater facilities in the hands of the manufacturer."[27]

## DECENTRALIZATION

While the central government has come to play an increasingly important role in British pollution control in the postwar period, the actual implementation of pollution-control policy remains relatively decentralized. Although local environmental health departments operate under legal authority granted to them by Parliament and the Department of the Environment, they continue to exercise substantial discretion, and in fact the attention local authorities devote to control of air pollution varies considerably from authority to authority. Michael Hill writes: "As far as day-to-day pollution control activities are concerned, central government merely gives local authority a frame-work within which there is considerable scope for creative implementation."[28]

The reluctance of the central government to impose national pollution-control requirements can be seen in the history of the implementation of the 1956 Clean Air Act.[29] Despite the clear threat to public health posed by the domestic burning of coal, the central government has generally refused to require local authorities to declare clean air areas (i.e., smokeless zones), even though it was explicitly given the power to do so by the Clean Air Act of 1968. Thus fifteen years after the passage of the original Clean Air Act, less than 10 percent of the acreage in forty-nine areas where the problems of smoke pollution were particularly acute was covered by smoke-control orders, and twenty-six local authorities had no acreage under control. Compliance follows a distinctly regional pattern: it is highest in London and lowest in the north of England and in the industrial Midlands. While authorities with serious problems of smoke pollution—the so-called black areas—are more likely to have established smokeless zones than those in "white" areas, the relative affluence of individual communities in the black areas explains much of the variance: the more affluent the community, the more willing its local government has been to pay its share of the costs of conversion. (The

1956 statute specified that 40 percent of the costs of converting the heating system of each residence and industrial site from coal to coke, oil, or natural gas would be borne by the central government, 30 percent by the local authority, and 30 percent by the occupant.)

An additional factor delaying full compliance has been the tradition in mining areas for miners to be given free coal as part of their pay; in the north of England, approximately 25 percent of the coal used for home heating is "concessionary" coal. While the National Coal Board has indicated its willingness to furnish a substitute cash payment, its offer has been viewed in many mining communities as insufficient to pay for the more expensive manufactured fuels. Moreover, many miners supplement their incomes by selling their concessionary coal. As a result, air quality has improved far more slowly in mining areas than in other parts of Great Britain.

Despite the efforts of the National Water Council during the 1970s to coordinate the activities of the regional water authorities and encourage them to establish and enforce their own water-quality objectives, the RWAs' enforcement efforts have varied considerably as a function of both the magnitude of the pollution problems they confront and their financial capacity to respond to them. An authority may, for example,

> set a highly restrictive consent with which it acknowledges, in private, the firm may be unable to comply.... This policy...enables discretion to be exercised over when and whether additional abatement is to be introduced by the firm...it enables the Authority to trade-off the implications of environmental damage on one hand, and the possible employment consequences of plant closure on the other.[30]

The standards imposed by the Alkali Inspectorate for particular processes have tended to be much more uniform in order to prevent plants in the same industry from securing a competitive advantage by meeting lower standards of pollution control. In applying the requirements published in its *Notes on Best Practicable Means* to any particular facility, however, the Inspectorate does consider local circumstances such as topography, prevailing winds, surrounding land uses, and existing pollution levels. Its 1973 annual report states:

> The Chief Inspector, with the help of his deputies, lays down the broad national policies and provided they keep within their broad lines, inspectors in the field have plenty of flexibility to take into account local circumstances and make suitable decisions. They are given plenty of autonomy and are trained as decision-makes with as much responsibility and authority as possible.[31]

Indeed, the very definition as to what constitutes an "infraction" of the Alkali Acts varies from inspector to inspector. The chief inspector noted in his 1951 annual report that there was "no well-defined dividing line between legal and illegal operation and a decision as to whether a given set of conditions shall or shall not be treated as an infraction rests, to a great extent, with the District Inspector."[32]

## RELATIONS WITH INDUSTRY

The British system of pollution control is predicated on a high degree of cooperation between the regulators and the regulated. Cooperation is particularly pronounced in the case of the Alkali Inspectorate, but it also characterizes the relationship between the regional water authorities and industry and, to only a slightly lesser extent, the interaction of local pollution-control authorities and industry; Timothy O'Riordan writes: "An important question in modern pollution control policy is the extent to which governments should force polluters to clean up. The traditional practice in the United Kingdom is to leave this matter largely to the good public sense of polluters in private consultation with regulatory officials."[33]

On balance, the ability of British pollution-control authorities to require compliance by industry is limited. British industry is under no statutory obligation to develop new technologies to control pollution, and government expenditures for this purpose have been limited in recent years. No law requires industry to keep the government informed of improvements in its technical ability to control its emissions, nor are firms under any obligation to share with government officials data on the costs of abatement. And if trade associations and individual companies do decide to disclose such information, the government's ability to assess its validity independently is, again, extremely limited. The Inspectorate, for example, employs neither economists nor accountants and must rely on an industry's own assessment of the state of its finances. Not only is the cooperation of industry thus critical to the government's ability to define what in fact constitutes the best practicable means of pollution control, but since much of the monitoring of industrial emissions is actually carried out by the companies themselves, British officials are also dependent on a high degree of voluntary compliance by industry with pollution-control requirements after they have been negotiated. While officials can prosecute companies for noncompliance, the financial penalties available to regulatory officials are fairly modest and pose no real economic threat to any but the smallest of plants.

By what means, then, do British pollution-control authorities secure the cooperation of industry? The most important is consultation with them. While government officials in Britain are under no legal obligation to consult with industry, in fact industry is closely consulted at every stage of regulatory policy, from the establishment of standards to their enforcement. The 1973 annual report of the Alkali Inspectorate describes the procedures by which the Inspectorate establishes its presumptive standards:

> Working parties and discussion groups are set up, consisting of representatives of the industry, its research organization, if any, and the Inspectorate.... The Inspectorate frequently travels abroad, sometimes in company with industry representatives, to examine foreign technology.... The chief inspector makes the final decision on any standards and other requirements... but this only follows mutual discussions with industry representatives.... Frequent, usually annual meetings are held between trade associations with the Inspectorate to note progress, discuss new technology, review research and development and generally reassess situations with the object of gaining further improvements and possibly getting together standard requirements for bpm.[34]

Similarly, the establishment of environmental quality objectives for each body of water by the regional water authorities takes place only after extensive consultations with each company that would be affected by them. (During the 1970s some members of each RWA were actually nominated by the Confederation of British Industry.) In addition, the Confederation of British Industry's Environmental and Technical Legislation Committee has established panels on clean air, synthetic detergents, water and effluents, industrial and solid wastes, and oil legislation, which work closely with government officials responsible for formulating policies in these areas.

Extensive interaction with industry also pervades the enforcement process: precisely because an important factor in the determination of bpm involves the costs of compliance for each individual source, pollution control is subject, of necessity, to continuous negotiations between inspectors and industry. These consultations can be understood in part as a strategy to secure compliance; by constantly consulting with industry, government officials in effect seek to co-opt industry into placing increased priority on pollution control. Thus "a substantial amount of a Field Officer's time is spent in creating and preserving good relations with dischargers of all kinds, even in the absence of pollution problems."[35]

The 1974 report of the Royal Commission on Environmental Pollution put the alternatives confronting British pollution-control officials in the following terms:

> Either we have, as now, an authority which because of its close relationship with industry and consequent understanding of the problems is able to assess the technical possibilities for improvement in detail and press for their adoption; or an authority which sees its job as one of imposing demands on industry and which, because of the sense of opposition that approach would create, could not obtain the same co-operation by industry in assessing the problems and devising solutions.[36]

The Commission added:

> The Inspectorate collaborates closely with industry in seeking solutions to pollution problems. The solutions which they eventually impose can be tougher and more practicable as a result of this involvement; the technical expertise of the Inspectorate is both essential to and fostered by this collaboration.[37]

The Inspectorate emphasized the critical role played by trade associations in compliance in its 1973 annual report:

> Participation by the trade associations [in setting standards] is a good guarantee of their support in enforcing requirements among the members, for they are anxious that all similar works in the country should have to meet the same requirements, with due allowance for adjustments to meet special local circumstances.[38]

Cooperation also plays an important role in the enforcement strategy of the regional water authorities:

> It ensures appreciation of the difficult technical problems in abatement incurred by companies. It enables the authority to be flexible in the setting of consents so that it can take into account the specific environmental circumstances of individual discharges.... The preferred procedure was for the technical officers of the company and the enforcement agency to discuss both the consent levels and the actual monitorings face to face ... thereby ensuring the type of compromise not possible in a formal courtroom setting.[39]

Cooperation between regulatory officials and industry is reinforced by the high degree of administrative flexibility built into the making and enforcement of pollution-control requirements. Because officials responsible for enforcement can negotiate agreements that will not—except in extremely unusual circumstances—be overruled by either their superiors or the courts, the British system is able to operate rather expeditiously. One official notes: "It is this ability to take responsibility and give quick decisions which pleases industry in its negotiations with the Inspectorate.... Industry will pay for time saved and the lack of 'red tape' and is prepared to accept tougher decisions than it might otherwise gain from protracted argument."[40] Moreover, the uncertainty that surrounds the definition of bpm contributes to a search for compromise, since neither side can be absolutely certain where the bpm boundary lies.

Cooperation is also reinforced by the considerable degree of mutual professional respect that characterizes the relationship between officials and industry. This is particularly pronounced in the case of the Alkali Inspectorate, whose members tend to be highly regarded for their technical competence. Inspectors, one official noted, are like "poachers turned policemen. We like to help industry [and] when we visit a works they are actually glad to see us."[41] The relations between industry and district health officials—who lack any specialized professional training in engineering—are somewhat more strained.

The chairman of the Chemical Industries Association Environment Committee informed a meeting of the Environmental Health Congress in 1981:

> In the UK, we are very fortunate in having relatively easy and frequent access to Civil Servants in both the Scientific and Administrative branches and many opportunities, both formal and informal, to state a case and to influence the opinions of policy-makers. By and large, our inspectors are practical, professional people, generally able, willing to take account of economic and employment factors so long as willingness to progress is being shown and progress made.[42]

The journal *Chemistry and Industry*, which regularly reviews the annual reports of the Inspectorate, echoed this appraisal:

> The Alkali Inspectors are a remarkable body of men. Any inspectors who are not looked upon by the inspected as having only nuisance value must be remarkable, and that is most certainly the case with the Alkali Inspectors. They have come to administrate an Act and a number of Orders which could easily become an intolerable nuisance to industry were they not intelligently and helpfully administrated.[43]

Regulatory officials tend to reciprocate these attitudes. A report on the Alkali Inspectorate notes:

The Inspectorate believes that "enforcement" as such is generally unnecessary; that industry is capable of regulating itself, and that a works that can prevent its emissions from causing a nuisance will do so without pressure. If there are repeated public complaints, the Inspectorate is likely to assume that a works has a technical problem and needs its help.[44]

A study of the regional water authorities reaches a similar conclusion: "Pollution control staff regard most dischargers as basically, if reluctantly, law-abiding...most are described as responsible or public-spirited people. Those polluters who cause serious difficulties...are a small minority."[45]

## PROSECUTION

Both underlying and reinforcing the cooperative relationship between regulatory officials and industry is the reluctance of the former to prosecute the latter. As one critic of British environmental policy notes, the Alkali Inspectorate "operates strictly within the canons of Good Form; of gradual reform rather than rapid change. It constantly displays a distaste for confrontation or any unpleasantness; it likes persuasion; it dislikes compulsion; it detests prosecution."[46] In the area of water pollution,

> modest (some would say feeble) levels of sanction and a marked unwillingness to prosecute are two of the major characteristics of regulatory law and practice.... A persistent failure to comply will be treated more seriously only when the discharge is regarded as substantially beyond consent limits and the pollution is noticeable.[47]

The Alkali Inspectorate has a similar policy: if suitable apologies are made in response to an official letter of notification of an infraction and steps taken to obviate recurrence, legal action is not sought.[48] Prosecution is regarded not only as reflecting badly on the company affected but also as a failure on the part of regulatory officials: it demonstrates that their efforts to persuade and educate industry as to their responsibility to control emissions have been unsuccessful. One water-pollution field officer remarked, "We don't take people to court just like that. It's a history of problems. We've tried everything with them: negotiations, discussion, etc. When we take them to court it's like saying all other methods have failed."[49]

During its first quarter century the Alkali Inspectorate brought only four prosecutions; its chief proudly informed the Royal Commission on Noxious Vapours in 1867, "I am rather proud of there having been so few prosecutions."[50] Between 1920 and 1966 only two firms were prosecuted. While the number of prosecutions increased during the 1970s—there were fifty-three between 1970 and 1978—they were directed primarily at small individual operations that had illegally burned cable, producing both dark smoke and hydrochloric acid. While violations of the Clean Air Act averaged 2,500 annually between 1970 and 1974, the number of prosecutions in England and Wales during this period ranged between 50 and 133 a year. Nor have local authorities been any less reluctant to prosecute individuals: between 1956 and 1960 there were only twenty-four prosecutions of individual households for violations of the Clean Air Act throughout all of Great Britain. And though the terms of many consents are violated, only 67 persons were prosecuted in 1971 under the terms of the Rivers (Prevention of Pollution) acts; in 1973 their number had increased to 137.[51]

The absence of legal sanctions reflects, in part, the ability of British regulatory officials to secure compliance through informal mechanisms of social control. A study of the enforcement activities of the regional water authorities reports:

> Much enforcement activity in pollution control work is premised on the assumption that discharges are suitable to extra-legal kinds of deterrence.... Few officers suggest that there is a stigma with economic implications reflected in damage to sales and profitability attaching to a manufacturer found guilty of polluting a watercourse. Rather the belief is that a company will seek to protect its reputation as a good in itself.[52]

Accordingly, "a common ploy" used by field officers "is to reason with their contacts, to appeal to their sense of social responsibility."[53] Explaining the lack of prosecutions by the Alkali Inspectorate, one official observed:

> Through the past hundred years...[enlightened managers] have gradually learnt to have confidence in the inspectorate's judgment when leading them along new paths and setting new and tougher standards....[By the end of World War II] industry had been thoroughly indoctrinated and inspectors were accepted as advisors.[54]

One inspector notes: "We come in as consultants rather than as stick-wavers. Then they feel obligated to do something...it's a matter of maneuvering them into a position where they have to act."[55] Moreover, Keith Hawkins reports:

> The more experienced field staff all detect a change toward greater pollution consciousness among dischargers and public alike that took place

88

in the 1950's and 1960's.... This is not to suggest that dischargers are incapable of deviance, but field staff believe that serious pollution that is a result of negligence or deliberate misconduct does not now occur regularly or on a large scale.[56]

British officials contend that an increase in prosecution would be counterproductive: it would undermine the cooperation on which the implementation of regulatory policy is based. As a recent chief inspector noted, "we look upon our job as educating industry, persuading it, cajoling it. We achieve far more this way. The Americans take a big stick and threaten, 'Solve your problem.' We say to industry, 'Look, lads, we've got a problem.' In this way we've got industry well and truly tamed."[57] An "assumption underlying compliance strategy is that dischargers are sensitive beings whose feelings may be easily bruised if urged or forced to do too much too soon. To use the big stick or crack the whip too zealously may well be counterproductive."[58]

The Royal Commission on Environmental Pollution makes essentially the same argument: "An aggressive policy of confrontation, involving prosecution for every lapse,... would harden attitudes and dispose industry to resist the imposition of costly programmes for pollution abatement."[59] A study of British water-pollution control policy reaches a similar conclusion:

> ...a policy of strict enforcement would destroy the spirit of co-operation with dischargers, painstakingly nurtured by Authorities for decades... although increased enforcement through the courts may yield short-term benefits of reduced discharges, in the longer term it could destroy firm–Authority cooperation, resulting in a decrease in information flow and an increase in the aggregate costs of reaching a specified water-quality level.[60]

Not surprisingly, the circular from the Ministry of Housing and Local Government that advised the water authorities on how to implement the 1951 legislation establishing a system of consents cautioned that "courts were to be used only as a last resort," adding that "since it was clear the greatest improvements in water quality were found where cooperation between discharger and authority was best developed, it was important to maintain good cooperation, which would not be cemented by frequent recourse to litigation."[61]

The lack of reliance on prosecution by pollution-control officials is both reflected in and reinforced by the government's policy of making fines relatively nominal for the majority of offenses. There is no provision for criminal prosecution under either the Alkali or Clean Air act; the maximum penalty is a fine of £400 with an additional £50 for each day on which an offense is repeated or continued after conviction. (In fact, the average fine on conviction is about £100, and fines of £30 or less are not uncommon.) For no apparent reason, the pollution of inland water or the contravention of a consent carries with it more severe penalties; under certain circumstances an unlimited fine, a term of imprisonment up to two years, or both may be imposed. In practice, however, the fines levied for most offenses by the regional water authorities are similar to those allowed under the legislation governing air pollution. As the "White Paper on Protection of the Environment" put it: the British system rests on "persuasion and the belief that, especially to industrial firms, it is the disgrace that counts, not the fine."[62]

## PESTICIDE REGULATION

A specific illustration of British reliance on voluntary compliance as a means of controlling pollution can be found in the area of pesticide regulation.[63] In the mid-1950s British officials became concerned about the effects of the growing use of pesticides on both human beings and animals. In 1959 an official working party under Sir (now Lord) Solly Zuckerman produced a report titled "Precautionary Measures against Toxic Chemicals Used in Agriculture." While it found no evidence that pesticides had become a danger, it did suggest that their use be carefully monitored. Following the report, the government appointed an interdepartmental advisory committee consisting of representatives of both industry and conservation groups. In the spring of 1957 a nonstatutory agreement was concluded between a number of government departments and the British Agricultural Chemical Association (BACA). It established a voluntary notification for pesticides known as the Pesticide Safety Precautions Scheme (PSPS).

The Pesticides Safety Precautions Scheme essentially represented a gentleman's agreement between industry and government, albeit with legal backing. The government assured companies that if they agreed to abide by the terms of PSPS, they would be considered in legal compliance. In turn the industry agreed to police itself. As an official of BACA put it, "If a company oversteps the mark, the rest of the industry is on the lookout.... They tell us and we tell the Government."[64]

Under the terms of this agreement, each manufacturer "voluntarily" agreed not to market any new pesticide that could cause new or increased risk without first securing the approval of the Advisory Committee on Pesticides (ACP). The ACP is officially described as "independent of commercial and sectional interests alike"; all of its members are scientists or doctors in government laboratories or universities.[65] In addition to determining the safety of pesticides for general use, the ACP recommends the precautions to be printed on the labels of new or significantly changed products. To assist the committee in conducting field trials and in monitoring pesticide use, a Wildlife Panel was established, consisting of two scientists nominated by the Inter-Departmental Advisory Committee, three from industry, and three from the Nature Conservancy Council, a statutory body established to promote and advise the government on wildlife conservation. Eight leading naturalist societies agreed to cooperate with the panel in conducting its research. The PSPS "plainly presupposed a climate of rationality, mutual trust and avoidance of emotional confrontation which contrasted sharply with the situation in the United States and elsewhere. [It represents] one of the first examples of detailed teamwork at all levels between government, industry and conservationists."[66]

Shortly after it was established, the committee faced its first major challenge: the Nature Conservancy reported a dramatic increase in the deaths of wild mammals and birds in areas where spraying was taking place. Among the animals killed were a considerable number of foxes. Immediately the masters of foxhounds, who preside over the still popular sport of fox hunting, became concerned. They began to pressure for an immediate investigation as to why more than a thousand of the animals they enjoyed pursuing across the open countryside had died prematurely. The chemical industry promised its full cooperation. A team of scientists from both industry and the Nature Conservancy, assisted by the government chemist and a team of veterinarians, began to trace the paths of toxic chemicals through the food chain in order to determine how pesticides and herbicides could be modified without impairment of their effectiveness. Shortly thereafter the Ministry of Agriculture announced that the manufacturers, the distributors, the farmers, the country landowners, and various scientific and conservation bodies had agreed to restrict the use of seed dressing that contained suspect chemicals. By 1970 many of the species that had suffered severe losses at the beginning of the 1960s were making an encouraging recovery, though problems still persist.

## PUBLIC PARTICIPATION

The close cooperation between industry and regulatory officials precludes opportunity for effective participation by other political constituencies. British pollution-control policy is basically made and enforced in private, by a "family-like, . . . close-knit group of experts proud of their traditions and the trust placed in them by the public."[67] O'Riordan writes:

> British environmental managers tend to feel that the public is passive and will accept what is thought good for it. . . . Most policymakers . . . see themselves as custodians of the public interest. . . . Regulatory policy making . . . [is] executed by selective consultation with particular interests but with no requirement to inform the general public.[68]

Both the negotiations between the Alkali Inspectorate and trade association officials to determine the best practicable means for each industrial process and the actual enforcement of the presumptive limits that are established take place strictly in private. (While the Trade Union Congress and the Institute of Environmental Health Officers are formally consulted with regard to the former, their role is a token one.) Nor is the public involved either in the negotiations on the terms of consents or in their enforcement by the regional water authorities. At the local level, environmental health departments are required to get the approval of committees of elected representatives before taking legal action against a company, but there is no requirement that these discussions take place in public. As a result, the opportunities that environmental pressure groups enjoy to influence the control of pollution in Britain are extremely limited.

The essentially paternalistic view of public administration that underlies the British system of pollution control is most clearly illustrated by the extremely limited information it makes available to the public. While the Alkali Act of 1906 did not require the Inspectorate to keep secret the data it collected on industrial emissions, the Inspectorate chose to abide by the highly restrictive terms of the Official Secret Acts of 1911. The Health and Safety at Work Act of 1974 did specifically forbid the Inspectorate to reveal the amount of pollution emitted by a firm under its jurisdiction unless the firm gave its consent. This change, however, merely gave legal recognition to what already was a well-established custom; the Inspectorate has historically viewed public disclosure as a threat to the willingness of factory managers to cooperate with it voluntarily. As the 1967 annual report of the Inspectorate put it, "abating air pollution is a technological problem—a matter for scientists and engineers."[69] The chief alkali inspector remarked in 1972, "I am a great believer in informing the public, but not in giving them figures they can't interpret. You would get amateur environmental experts and university scientists playing around with them. People can become scared of figures, they can get the wind up."[70] The secrecy surrounding the

Inspectorate's efforts extends to its enforcement activities: it does not disclose the names of firms found guilty of willful or careless pollution. Their names appear in the Inspectorate's annual report disguised by such terms as "one metal recovery work" and "a tar work."

The activities of other pollution-control authorities are similarly secret. Under section 7 of the Rivers (Prevention of Pollution) Act of 1951, river authorities were required to maintain a register of consents as well as the conditions required for them and to keep it "open to inspection at all reasonable hours by any person appearing to the river board to be interested in the outlet."[71] The definition of the term "interested," however, is up to each authority, and virtually all have interpreted it in a highly restrictive manner, usually limiting it to "those with a property interest at the point of discharge"; that is, the polluter. The Rivers (Prevention of Pollution) Act of 1961 subsequently made it a criminal offense for the water authorities to disclose information on pollution which had been obtained either in connection with an application for consent or in the course of an inspection. An exception could be made only if the firm that supplied the information or produced the pollution gave its permission. Section 26 of the Clean Air Act of 1956 as well as section 59 of the Offices, Shops, and Railway Premises Act of 1963 threatened inspectors from local authorities with three months' imprisonment for the unauthorized disclosure of information, thus making the penalties for the disclosure of the law's enforcement more severe than those for violations of its provisions.

The control of pesticides follows a similar pattern. Policy is made by a group of technical experts, with virtually no opportunity for public participation.

> PSPS advisors have developed a long tradition of working closely with pesticide manufacturers and advising a government department with a predominant interest in the promotion of agricultural production. Neither of the groups with a direct interest in the regulation of potentially carcinogenic substances—workers and consumers—are represented in PSPS's decision-making process.[72]

Between 1957 and 1977 the Advisory Committee on Pesticides published only five reports describing the toxicity studies on pesticides that it had received. Forty-eight applications for new products or uses were turned down by the PSPS between 1973 and 1977, but the Ministry of Agriculture refused to disclose the names of the pesticides that failed to obtain clearance.

## PLANNING AND POLLUTION

One vehicle does exist by which individuals and community groups can attempt to control the emissions to which they are exposed. They can use their power under the Town and Country Planning Act either to deny approval for developments likely to increase pollution or to insist that new investments meet certain conditions with respect to environmental quality. Following the increase in public concern about pollution in the late 1960s, the Town and Country Planning Act was amended in 1971 to require local planning authorities to formulate "policy and general proposals in respect of the development and other use of land in that area, [including] measures for the improvement of the physical environment." Such measures clearly encompassed attempts to control pollution. The Department of the Environment subsequently instructed local authorities that planning "proposals should be related to the need to combat and prevent pollution" and urged that "pollution policies should be embodied in [local] structure plans."[73] A Town and Country Planning development order issued the following year required local planning authorities to consult regional water authorities before granting permission for certain kinds of developments. As a result, local planning authorities have become more aware of their responsibilities for controlling pollution. Between 1974 and 1980 approximately a third of all district planning authorities in England and Wales used their planning powers—most commonly via the imposition of consents—to control air pollution and nearly half had used their planning powers to control water quality.[74]

Despite the increased use of controls over development as a device for controlling pollution, local planning authorities who wish to impose pollution controls on industry face many obstacles. First, because the British government specifically discourages the use of environmental quality standards, local planning authorities have "no explicit objectives or policies related to pollution."[75] Some local governments, most notably the Greater London Council, have made reference to air-quality standards in their development plans, but when one authority, the Cheshire County Council, sought to include within its structure plan a prohibition on any industrial expansion that would produce concentrations of smoke and sulfur dioxide in excess of World Health Organization standards, its plan was denied approval by the secretary of state for the environment.[76]

In addition, local planning authorities are in a position to exercise control over pollution at only one point in time: when an application is considered. Local authorities can attempt to revoke permission or modify the conditions attached to it after it has been approved, but these procedures require the approval of the secretary of state for the environment and it is rarely granted. As a result, once planning permission has been granted, it is extremely rare for local planning authorities to take legal action to ensure that their consent conditions are fulfilled.

A more important constraint confronted by local planning authorities lies in the conflict between their responsibilities to protect the environment and those exercised by the pollution-control officials of the central government. The Department of the Environment has resisted most attempts by local planning authorities to displace the role of the Alkali Inspectorate by establishing consents for registered firms that exceed those required by the Inspectorate. Accordingly, only a small portion of the conditions attached by local authorities apply to plants registered by the Inspectorate. While there is frequent consultation between local planning officials and the Inspectorate, on a number of occasions the Inspectorate has taken the part of industry in opposing planning conditions urged by local planning authorities on the grounds that they either were unnecessary or duplicated its own enforcement efforts. For the most part, the Department of the Environment has sided with the Inspectorate against local planning authorities. In its circular issued in 1977, the Department of the Environment warned planning authorities that they

> should work in close liaison with those responsible for pollution control and the disposal of waste. Planning authorities are reminded...that they should exercise care in considering the use of conditional planning permission....Planning conditions should not be imposed in an attempt to deal with problems which are the subject of controls under separate environmental legislation, whether the controlling authority belongs to central or local government.[77]

Even the Royal Commission specially advised planning authorities that if they were

> using [the proposed] air quality guidelines and concluded that an unacceptable amount of pollution is likely to be emitted from a proposed plant when the Alkali Inspectorate's requirements have been met, their sanction should be the refusal of planning permission, not the imposition of planning conditions designed to control emissions.[78]

This alternative is often impracticable, however, since local planning authorities are generally reluctant to deny planning permission altogether because of the economic consequences of doing so. Despite these limitations, the Town and Country Planning system has provided an important arena for public participation in the shaping of pollution-control policy in Britain—a subject that will be explored in more detail in Chapter 3.

## PRESSURES FOR REFORM

The secrecy surrounding the control of pollution emerged as a political issue in the early 1970s. In 1972 a journalist, Jeremy Bugler, published a popular book, *Polluting Britain,* in which he strongly attacked both the secrecy of the Alkali Inspectorate and the closeness of its ties with industry.[79] That same year the Royal Commission on Environmental Pollution issued a report that "strongly supported the citizen's right to know what pollutants were being put into air and water or dumped on land" and urged that the Department of the Environment officially recognize this right.[80] A year later Lord Ashby, the first chairman of the Royal Commission, told the House of Lords that the "miasma of confidentiality" surrounding the Inspectorate ought to be removed.[81]

In 1972 a bill was submitted to Parliament to require the Alkali Inspectorate to consult with local authorities before establishing emission standards for works in their area; not only would these standards be published, but "any record, report or information" that did not involve genuine trade secrets "would also be publicly disclosed."[82] In submitting his bill, Mr. Neil McBride, the MP for Swansea, cited the "mounting criticism" of the Inspectorate and the "cloak of anonymity" it had worn since the 1860s, and concluded that it was high time it "emerge[d] from the Victorian shadows where it has been lurking willingly for far too long."[83] While this specific proposal was not taken seriously, in the years both before and preceding its submission various government ministers found themselves the subjects of persistent criticism from members of Parliament. They raised such questions as:

> Why were so few industrial polluters prosecuted? Why, on the rare occasions when there were convictions, were the fines so derisory? Why were the inspectors so secretive about the data they collected and the standards they prescribed? What instructions were given to the Inspectorate about the interpretation of best practicable means?[84]

Writing in the March 9, 1972, issue of *New Scientist,* Jon Tinker strongly criticized the "nanny knows best" mentality that appeared to underlie the attitude of the British government toward increased public knowledge of its pollution-control efforts. He asked:

Is flexible, cooperative British pollution control the best in the world, as many of its practitioners claim? Or do flexible officials sometimes place public health and safety at risk rather than jeopardize their cooperative relationship with polluters? The public is unable to judge, because the facts which they need to inform an opinion are hidden away in secret cabinets, guarded by the cooperative ranks of industry and its pollution officers, protected by the majesty of a dozen acts of Parliament. . . . In Britain today, data on individual emissions are guarded more closely than military blueprints.[85]

A director of the Friends of the Earth had reached a similar conclusion: "Everybody *says* [the inspectors are] very good, but how can we tell if they don't publish the figures?" Tinker argued: "The notion that government alone is clever enough to evaluate pollution problems, and that the general public is too ignorant and too stupid to understand them, I find paternalistic and offensive."[86] And in 1974 a special report of the Social Audit, a public interest research group, concluded that the Inspectorate can be made "fully accountable" only "if the secrecy which now protects industry is replaced by complete and honest information to the public."[87]

The Control of Pollution Act (1974) was in part a response to these criticisms. While its primary purpose was to consolidate several previous pollution-control statutes, some of its provisions did make pollution-control policy in Britain more public. The statute required that all consents to discharge wastes onto the land, into the water, and into the air be published in a public register, subject only to appeal to the secretary of state for the environment on the ground of confidentiality. It also required each RWA to disclose its own sewer discharges. These requirements are extremely significant since once consents were published and actual discharges monitored, a citizen could sue if a consent had been exceeded—even though he or she would still have no right to challenge the terms of the consents themselves. As a result, it would become

much easier for clubs and conservation societies, as well as individual property owners, to bring civil actions against polluters. . . . It will also enable such persons to exert pressure on water authorities to take appropriate action, and could even lead to private prosecutions of discharges, including a water authority itself.[88]

For the next ten years each environment secretary chose to delay the implementation of these sections of the 1974 legislation. The government's policy officially reflected its reluctance to impose additional administrative costs on both the RWAs and local authorities. But there was also an additional factor: a disproportionate share of the plants discharging effluents in excess of existing consents were municipal sewage treatment facilities. The administrative reorganization of the British water system had not led to the improvement in British water quality that was expected for the simple reason that the sewerage works inherited by the regional water authorities were totally inadequate to cope with the volume of effluent produced by British industry. Although the real value of investment in sewage treatment increased fivefold between 1951 and 1971, the RWAs found themselves the greatest abusers of their own standards. Moreover, their efforts to update the sewerage facilities of England and Wales were severely hampered by Treasury restrictions imposed in the early 1970s on additional borrowing by nationalized industries. (In fact, a major share of the expenditures of many RWAs involves interest payments on their debts.)

Although the government finally announced in 1984 that Part II of the Control of Pollution Act would be phased in over the next four years, it is not obvious that this action will result in greater public accountability on the part of the regional water authorities. For in 1977, anticipating the eventual implementation of the 1974 statute, the National Water Council began a systematic review of existing consents throughout England and Scotland. Its stated purpose was to enable Britain to obtain the "best environmental value" from the limited resources the nation had available to devote to improving water quality. The council was particularly concerned that the majority of consent conditions that had been adopted by the predecessors of the RWAs relied too heavily on the Royal Commission's biological oxygen demand standard, which had been established in 1912, and that the result could be "wasteful allocation of resources in maintaining or upgrading sewage treatment capacity and standards to meet consents where higher quality effluent was not strictly required by use of the receiving watercourse."[89] In fact, as the National Anglers' Council and others have argued, the real purpose of the review appeared to be to bring consents more in line with actual practices, thus reducing the vulnerability of both polluting companies and the RWAs to lawsuits if and when the disclosure provisions of the 1974 statutes were implemented.

The Control of Pollution Act of 1974 also attempted to enlarge the role of local authorities—who generally tend to be more responsive to complaints about emissions than officials of the central government—in the monitoring of air pollution. Sections 79 to 83 increased the ability of local authorities to investigate air pollution by enabling them either to require emissions data from local firms or to make their own measurements. Any information they obtain under the act must be made public unless the secretary of state determines that publication would either jeopardize a trade secret or be "contrary to the public interest," or that "information cannot be provided except at undue expense." If a local authority chooses to take advantage of these powers, however, it must, according to a government circular published in 1977 when Part IV of the Control of Pollution Act came in force, establish a local consultative committee containing "equal numbers of local authority members and representatives of local industry."[90]

This provision, however, has not resulted in furnishing the public with much new information concerning emissions for the simple reason that very few, if any, local authorities have chosen to implement it. Michael Hill, who has closely studied the role of British local governments in pollution control, concludes, "Those about which I have secured details have been satisfied so far to obtain the voluntary cooperation of industrialists and have not used their *powers* to obtain information.... Many authorities would appear to have given no serious attention to this legislation." One chief environmental health officer, Hill reports, told a conference dealing with this subject that "if there had been pollution problems, local authorities probably had ways of finding this out and collaborating with industry. This is the case in Bristol; therefore we want to avoid additional consultations and meetings if we are not going to get much out of them." Hill adds:

> Since much of their other pollution control work rests...on the development of a cooperative relationship with polluters, it is inevitably difficult for local authorities to force the issue of disclosure, particularly as their opponents can argue that the results of such an activity could well be of little practical value but raise public anxiety for the local authority to handle.[91]

In its Fifth Report, published in 1976, the Royal Commission on Environmental Pollution noted that "many people are suspicious of the Inspectorate's relationship with industry and believe that a more aggressive attitude is necessary if pollution problems are to be attacked with sufficient vigor. It is commonly remarked by critics that this relationship is 'cosy.'" While contending that its own investigation of the Inspectorate's enforcement policies did not reveal this to be the case—indeed, the commission was impressed by the toughness of the Inspectorate's bargaining with industry—it did suggest that "a system operating on the basis of co-operation with industry has a particular need to satisfy the public and especially communities seriously affected by pollution, that it is sufficiently firm in pressing for improvement." The commission noted that the Inspectorate appeared to have made little effort to respond to the public's increased interest in pollution control; many inspectors continued to regard any questioning of their judgments by the public as presumptuous and saw no reason to justify or explain their policies to the public as a whole. It concluded:

> This is not to say that the agreements reached between the Inspectorate and industry about abatement measures are unsound.... Nevertheless... it is no longer acceptable that decisions on emissions which directly affect the daily lives of many people should be taken by a small, specialist body consulting only with industry; greater participation is needed, not least so that the assumptions and problems on which the decisions depend are more widely understood.[92]

Accordingly, the commission urged that the Inspectorate seek a change in the Control of Pollution Act of 1974 that would allow it to release to the public the emissions data supplied to it by industry. It also suggested that the Inspectorate establish a system of formal consents similar to that employed by the regional water authorities and that it make public the terms of these consents, as well as industry's compliance with them. While the commission did not want to undermine the flexibility associated with the use of "best practicable means," it did urge that the public be supplied with more information about how they were determined. The commission firmly rejected the establishment of strict, legally enforceable emissions standards on the grounds that they would encourage confrontation with industry without reducing pollution levels, but it did propose the establishment of air-quality guidelines for five pollutants in order to provide both industry and the public with a clear indication "as to whether or not the air at a particular time and place should be considered unacceptably polluted."[93] To date, none of those recommendations has been adopted.

Nevertheless, the amount of information made available to the public has gradually increased. In its Tenth Report, published in 1984, the Royal Commission noted that "there is a growing acceptance on the part of industry and pollution control authorities of the need to share information with the public."[94] A number of water authorities, even before the official implementation of the 1974 disclosure provisions, have released details about both their consents and their monitoring of water quality. These disclosures have been made with the permission of the firms under their jurisdiction. The Alkali Inspectorate, for its part, has disclosed to local authorities the details of bpm requirements for individual plants and also has made the annual reports of each district inspector available to local authorities, plant managers, and interested members of the public. The Inspectorate has encouraged the establishment of local liaison committees to discuss problems related to specific plants.

On balance, the British public seems to be relatively satisfied by the efforts of its government to improve air and water quality. Some communities have expressed considerable resentment at the inability or unwillingness of the Alkali Inspectorate to curb the emissions from a neighboring plant; on one occasion in the early 1970s housewives blockaded the entrances and exits of a local plant to protest the unresponsiveness of the Inspectorate to their complaints about its emissions. The highly visible pollution stemming from the Bedfordshire brickworks has long been the focus of considerable public criticism of the Inspectorate's enforcement efforts, and the pollution-control efforts of various RWAs have also been strongly criticized.

But while Tinker correctly observes that "the Conservation Society, the Friends of the Earth, and other young eco-action clubs tend to start from the assumption that, as has proved to be the case in the USA, government pollution agencies are normally incompetent and hand-in-glove with industry," these sentiments are not widely shared even within the British environmental movement.[95] There is in Britain today no significant domestic pressure to change the way British pollution-control policy is either made or enforced. Complaints about pollution tend to focus on particular sources, not on the system of regulation itself. Within the context of British environmental policy as a whole, conservation and land-use issues loom far more important and appear much more capable of mobilizing public opinion.

This lack of conflict is due in part to the low political visibility of most British pollution-control officials. In a study comparing the control of air pollution in the United States and Britain, Christopher Wood writes: "The Alkali Inspectorate has sometimes been attacked as being industry's ally, secretive and remote, but these accusations have not been widely supported: most people in the United Kingdom have never heard of the Alkali Inspectorate." He adds: "Control by local authorities under the Clean Air Acts and the Public Health Act is even less controversial."[96] Moreover, because responsibilities for controlling air pollution are split between the Alkali Inspectorate and the Environmental Health Department, the average citizen is likely to be uncertain as to whom to address complaints about air quality, and in any case, few know how to find the Alkali Inspectorate. Nor, for the most part, does the public appear to be aware of the pollution-control policies of the various regional water authorities. The decentralization built into the British system of pollution control also serves to diffuse pressures for change: criticisms of the abatement practices of particular works tend to remain local in scope.

There is an additional reason for the relative lack of political controversy surrounding British pollution-control policy: over the last three decades considerable progress has in fact been made in improving both air and water quality. A. H. Perry writes: "In general air pollution is not regarded as a serious problem by the public and there is little public anxiety about it, perhaps because the single most visible component of air pollution—smoke—has declined."[97] There has been an equally substantial and highly visible improvement in the quality of the Thames, Britain's most visible river. These accomplishments are not only recognized by the British public but even acknowledged by many of the system's most vocal critics. Jeremy Bugler, while highly suspicious of the close ties between the Inspectorate and industry, nonetheless concedes that "the British industrial air-pollution picture is not without its achievements." He adds: "Clearly it would be absurd to credit all these reductions to the visits of district Alkali Inspectors complete with briefcases and good manners. But some credit is due."[98]

## THE INTERNATIONAL CONTEXT

Since Britain's entry into the European Community in 1973, the contrast between the British and continental approaches to pollution control has become a source of considerable political tension.[99] The same year that Britain joined the European Common Market, an environmental program was formally adopted by the Community's Council of Ministers. Taking as its basis article 2 of the Treaty of Rome, which affirms that an objective of the Community is to promote "a harmonious development of economic activities," the EEC's "action plan" states that "major aspects of environmental policy in individual countries must no longer be planned or implemented in isolation. On the basis of a common long-term concept, national programmes in these fields should be coordinated and national policies should be harmonized within the community."[100] In practice, "harmonization" has meant not only the establishment of uniform emissions and air- and water-quality standards throughout the member nations of the Community but also the standardization of pollution-control procedures. For its part, the EEC's efforts to implement this program reflect in large measure not only the extremely serious pollution problem on many parts of the Continent but also its international nature, particularly in the case of water pollution. Officials in Brussels contend that unless strict, uniform standards are established, no member of the Community will match the pollution-control efforts of its neighbors. In particular, Community officials are concerned that Britain's lack of fixed standards makes it impossible for the EEC to assess whether in fact Britain is complying with its directives.

Britain's strong resistance to the commission's initiatives in this area stems in part from its officials' desire to maintain their nation's distinctive approach to pollution control—one in which both industry and government take a great deal of pride and satisfaction. But there is also a more pragmatic consideration. Since the capacity of much of the British environment to absorb relatively large quantities of pollutants without adverse affects on health or amenity is substantially greater than that of the Continent, any effort to establish uniform emission standards would result in greater expenditures on the part of British industry than appear to be warranted by the actual extent of Britain's pollution problems. As one British official put it, "Italy economically benefits from the amount of sunshine it receives each year. Why should not our industry be able to take similar advantage of our long coastline, high winds and rapidly flowing rivers?" A British industrialist remarks:

> In favor of uniform national and international standards it is sometimes argued that the latter system avoids giving an unfair commercial advantage to industries in locations where the environment can absorb and deal with more pollutants than in others. This argument is just as logical as claiming that all factories, whatever their locations and distance from markets or sources of raw materials, should carry the same transport charges.[101]

Writing in the *Review* of the Confederation of British Industry (CBI), Eric Felgate argued that

> emission standards for industries in countries so widely separated as the ten nations of the enlarged Community cannot sensibly be made uniform, any more than they can within the UK itself. Emission standards must have regard to local conditions and circumstances: if they are uniform, they will invariably be found too restrictive in some areas, too lax in others.[102]

In response to Community pressures, however, Britain has made some changes in its approach to pollution control. Specifically, it has adopted Community directives establishing air-quality standards for three pollutants: sulfur dioxide, particulates, and lead. It has also agreed to a Community directive specifying upper and lower limits to the amount of lead in petrol, and has agreed to abide by a directive issued on March 4, 1976, requiring member states to adopt uniform emission standards for five particularly dangerous substances—mercury, cadmium, aldrin, dieldrin, and endrin—after it was amended to allow Britain to substitute quality controls (which measure average pollution levels) for strict emissions standards (which measure actual emissions). While the effects of these changes remain marginal, the British are finding themselves under continual pressure to bring their system of pollution control more in line with that of the other members of the European Community. As one executive noted, "it may well be that the British system, like some wines, will not travel, which is all the more reason why the CBI will be keeping a close watch on developments in community policy."[103]

Britain has also found itself under strong international pressure to restrict its emissions of sulfur dioxide, particularly those that leave its borders. Since the mid-1950s the acidity of rain over parts of Scandinavia has substantially increased: according to one Swedish scientist, concentrations of hydrogen ions increased forty-fold between 1955 and 1968. These changes in precipitation have caused particular concern in Sweden, where the increasing acidity of rain has been blamed for a reduction in forestry yields and a decline in the number of fish. In 1972 the OECD undertook a study of the long-range transport of air pollutants within Western Europe. It found that "imported" pollutants accounted for 77 percent of the sulfur deposited in Norway's atmosphere, 70 percent of that deposited in Sweden. Britain was found to be among the leading exporters of sulfur dioxide to both countries, accounting for between 11 and 16 percent of sulfur pollution in Sweden and between 26 and 47 percent of sulfur pollution in Norway. Between 1973 and 1974 approximately 60 percent of the sulfur dioxide put into the air in Britain was carried eastward by prevailing winds.[104]

Britain is both the largest producer and the largest exporter of sulfur dioxide in Europe. Its production of sulfur is due to its dependence on coal as a source of fuel: in comparison with both the Federal Republic of Germany and France, a relatively small proportion of Britain's power is produced by nuclear energy. The strategy adopted by the Alkali Inspectorate to control emissions from coal-burning power stations has involved primarily the construction of tall stacks—approximately 200 meters high—that shoot emissions upward into the atmosphere at high pressure; all large power plants in Britain employ this method of pollution control. The Alkali Inspectorate has contended that this strategy is the best practicable means of pollution control and is consistent with the control strategy adopted by other industrial nations, including Sweden and the United States. (While the EPA has also promoted the installation of scrubbers, only 10 percent of American power plants have actually installed them.) Throughout the 1970s, officials of the Central Electricity Generating Board (CEGB), whose power plants are responsible for half of Britain's sulfur dioxide emissions, strongly resisted the installation of scrubbers that would reduce the sulfur content of coal before it is burned. They argued not only that scrubbers would add significantly to the cost of constructing new power stations but that they "cause more problems than they relieve, displacing relatively harmless gaseous waste with more problematic solid wastes."[105] Their position was supported by the Alkali Inspectorate, which refused to require the installation of scrubbers on the grounds of costs, operating problems, the difficulty of disposing of the sludge, and the uncertain magnitude of the benefits of control.

The British position appeared to change somewhat in 1981. The Alkali Inspectorate noted in its annual report that scrubbers appeared to have developed into a feasible means of pollution control, capable of reducing sulfur dioxide emissions by 90 percent while adding only 10 to 20 percent to the cost of power plant construction. It also observed that scrubbers were being installed on an increasing number of plants in both the United States and Japan. For the first time the report expressed official interest in other advanced technological approaches to pollution control, including oil desulfurization, fluidized bed combustion, and improved coal-cleaning techniques. While not officially changing its *Notes on Best Practicable Means* for new coal-fired power plants, the Inspectorate did suggest that such a change was likely "in light of experience now being gained in other countries."[106] But since no new coal-fired plants are currently being planned for the United Kingdom—the emphasis now is on the construction of nuclear facilities—this policy shift is largely symbolic.

While power stations built in recent years have been required to leave sufficient space for the installation of scrubbers, both government and industry remain strongly opposed to the retrofitting of Britain's existing coal-burning factories. The CEGB, for its part, remains "proud of Britain's historically advanced pollution control programs" and unenthusiastic about the installation of additional controls. As one of their spokesmen recently put it, "the United Kingdom was the first country to become polluted and also the first country to clean up its pollution. Particulate pollutants have been reduced by more than 80 percent, and sulfur pollutants have been reduced by more than 50 percent. We believe we have behaved very responsibly in this area."[107] In fact, Britain's sulfur dioxide emissions declined by 37 percent between 1970 and 1980. Britain, however, has refused to join the "30 percent club," a group of European nations that have pledged to reduce sulfur emissions by 30 percent between 1980 and 1993, on the grounds that such a commitment would add about 5 percent to the nation's electricity bill; it has, however, agreed to aim for a 30 percent reduction by the end of the century. The government's decision was strongly criticized in a report issued by a select committee of the House of Commons in December 1984, which warned that acid rain was placing much of Britain's countryside at risk as well.

Aside from the issue of cost, both the CEGB and the Alkali Inspectorate contend that scientific evidence linking Britain's emissions of sulfur dioxide to the deterioration of Sweden's forests and lakes is by no means conclusive. They argue that the productivity of a nation's forests is affected by a wide variety of environmental and climatic factors and that the relative importance of acid rain has not been clearly established. British scientists also point out that the salmon catches in some Swedish rivers were declining long before acid rain became a problem and that "the possible effects of acid rain have still to be disentangled from changing practices of land-use (affecting run-off into rivers), epidemic disease, and over-fishing."[108] They have also suggested that nitrous oxides, much of which are emitted by automobiles, may be as great a cause of acid rain as sulfuric acid. While acknowledging that "acid rain will be seen as a major and pressing international issue over the next few years," the Alkali Inspectorate insists that "less is known about the effectiveness of remedial measures and there is doubt that widespread and expensive measures to reduce sulfur emissions (even if this were possible) would result in a significant improvement in the aquatic environment."[109] The CEGB has, however, commissioned investigations into the processes by which sulfur dioxide emissions are transformed to acid sulfates in the atmosphere. This controversy will doubtless continue.

## CONCLUSION

Over the last century Britain has evolved a distinctive approach to the control of industrial emissions. Its distinction lies less in its restrictions on nonindustry participation in the regulatory process or in the degree of cooperation between industry and government—both of which also characterize the regulatory systems of other nations—than in its informality and flexibility. In sharp contrast to the approach adopted on the Continent and in Japan, Britain's strategy for controlling emissions remains rooted in the traditions of the common law. In response to both international and domestic pressures, pollution control in Britain has undergone numerous changes over the last two decades: it has become more centralized, more public, more legalistic, and in some respects more contentious. The pace of change remains modest, however: the fundamental principles on which pollution control in Britain rests have not changed significantly over the last century.

# CHAPTER THREE

# The Politics and Administration of Land-Use Planning

Land-use planning in Britain differs in several respects from the control of pollution. While emissions are regulated in a highly secretive manner, decisions regarding land use are highly visible. The system of public inquiries that constitutes the political focus of British planning policy provides considerable opportunities for participation by both amenity groups and community residents. Precisely because the British system of pollution control is so closed, a disproportionate amount of the conflict surrounding British environmental policy has centered on decisions affecting land use. In comparison with pollution-control regulation, British planning policy has been fairly contentious: particular land-use decisions, the procedures by which they have been made, and even the scope of the planning system itself have been the focus of considerable controversy. Moreover, the British public seems to be much more concerned about conservation issues than about the control of pollution. As we noted in Chapter 1, Britain has a highly organized network of conservation organizations and they have exerted an important influence on public policy at both the national and local levels.

## THE PLANNING SYSTEM

### Procedures

The origins of British planning date from 1890, when local authorities were given the power to remove slums and construct new housing. Two decades later Parliament enacted the Housing, Town Planning, Etc. Act, which recognized that "amenity" and "the laying out and use of any neighboring land" were factors to be considered in planning "schemes" for land about to be developed.[1] Since the 1920s a succession of Town and Country Planning acts—the most important of which was enacted in 1947 and the most recent in 1980—have progressively expanded the scope of the powers exercised both by the secretary of state for the environment and by local planning authorities. Currently both county council and county borough council planning authorities are required to develop plans that, subject to ministerial approval, provide a framework within which specific proposals for development can be assessed. These plans frequently include strategies for dealing with a wide range of social questions, including employment, recreation, population, and transportation.

The provisions of the Town and Country Planning Act of 1971 provide that if the character of a proposed development would be likely to create a nuisance under the common law, affect the character or appearance of a conservation area, or conflict with the implementation of the decisions of a development plan already in effect, then the local authority is under both a statutory and moral obligation to publicize it adequately and to invite any interested parties to submit their views.[2] While this requirement applies only to developments proposed by private individuals or firms, in practice it applies to most public-sector developments as well. The local planning authority is required by law to make a decision within eight weeks of receiving the initial application. It can either reject the application or accept it, either as submitted or subject to various conditions. If it accepts an application, provided proper procedures, including any requiring public notification, have been followed, its decision is final.

If, however, the planning authority decides to reject the application or insists on conditions unacceptable to the developer, the latter has the right to appeal to the appropriate minister—the secretary of state for the environment for sites in England, the secretary of state for Scotland for sites in Scotland, and the secretary of state for Wales for sites in Wales. In such cases, the appropriate minister, in order to provide himself "with the information he needs in order to arrive at a decision on the merits of an application," will either invite the interested parties to submit written statements or authorize an inspector to hold a local public inquiry.[3] Alternatively, if a proposal is regarded as having more than local significance, the minister may choose to "call in" the application as soon as it is submitted, thus bypassing the local planning authority's decision role.

The local public inquiry represents, in essence, the modern counterpart of the hearings that were held more than a century ago at which property owners were given the opportunity to challenge the government's compulsory purchase of their land. A former secretary of state for the environment has described the public inquiry in the following terms:

> I believe that as a result of our inquiry processes we can make better informed decisions which fit the facts and which fit into national, regional and local objectives; second, our inquiry system enables us to hear and to have answered the legitimate anxieties of people who have the right to express their concern; third, the public examination of these large questions assists us in achieving a measure of consent when final decisions are made.[4]

Public inquiries are conducted by a corps of some 200 inspectors, nearly all of whom are professional planners, engineers, or architects. While both proponents and opponents of a development scheme are given the opportunity to cross-examine each other's witnesses, hearings tend to be more informal than judicial proceedings. There are no formal rules of evidence and individuals need not be represented by counsel—though larger commercial interests usually are. Most important, local residents, whether speaking as individuals or as representatives of voluntary associations, are encouraged to participate and communicate their views.

At the close of the inquiry the inspector, accompanied by the parties involved, personally visits the site of the proposed development in order to judge for himself its probable effect on the community. Then, on the basis of what he has seen and heard, he writes a report that usually culminates in a recommendation. The processing of the inspector's recommendation depends on the nature of the controversy. A straightforward case is generally decided by either a senior executive officer or a principal officer, while more complex cases or those that raise issues of national importance are resolved at a higher level in the departmental hierarchy, often involving the parliamentary or assistant secretary. While ministers are under no obligation to follow an inspector's recommendation, "there is a strong presumption in favor of his judgment, and it usually requires a consensus of opinion among the most senior and experienced officials in the department before any question of rejecting his recommendation arises." In most cases these officials simply decide in the name of the minister. However, in cases that involve considerable controversy and may have attracted the notice of Parliament, that raise a new issue not covered by existing policy, or that simply are so unusual that the minister's personal judgment is required, the final decision is made by the minister himself, or one of his junior ministers. In general, "when civil servants feel they know what decision the minister would take, they take it themselves. When they are unsure, they do not." As one civil servant put it, "whatever he wants to see, he sees and whatever we think he ought to see, again he sees."[5]

Although Britain's three secretaries of state have the final say in determining land-use decisions, in fact the British planning policy remains highly decentralized. Only 5 percent of the decisions of local planning authorities are appealed and an even smaller number are "called in." While the circulars issued by the Department of the Environment do play an important role in shaping local planning decisions, each local planning authority does in practice enjoy considerable autonomy.

A second mechanism by which planning controversies may be resolved is a private act of Parliament. At the beginning of the Industrial Revolution, any developer, either private or public, who wished to build on land that he did not own was required to secure the approval of Parliament. He did so by means of a private act of Parliament, a law whose purpose was to confer "an interest or benefit on a particular person or body of persons rather than on the public in general."[6] (Much of the enclosure movement worked through this legal device.) Over the last 150 years Parliament has successively delegated more and more of its powers to control land use to ministers and lesser public authorities. Some developments, however—usually those involving the compulsory purchase of land by such public authorities as the RWAs—continue to proceed by private bills.

The procedure followed by Parliament in such cases is more formal than that of a public inquiry. A small panel of MPs or peers conducts a hearing at which both the developer and people opposed to the development are customarily represented by barristers who specialize in parliamentary private business. Since the petitioners (those opposed to the development) are in effect alleging that the promoters are inflicting an injury on their interest, the committee hearing bears an important resemblance to a court action. Once a bill has received committee approval, it is normally difficult for petitioners to generate sufficient interest among MPs to oppose its passage, though on occasion committee recommendations have been vigorously debated by the full house.[7] If they lose in the House of Commons, objectors have another opportunity in the House of Lords, where the influence of environmentalists is often greater. The use of a private act of Parliament as an alternative to planning permission is relatively rare, however, and almost always involves developments sponsored by a nationalized firm, such as a regional water authority, or a statutory undertaker, such as the National Coal Board.

*Administrative Discretion*

Despite the existence of carefully defined procedures for handling planning applications, the British planning system still leaves substantial room for administrative discretion. The central government is under no obligation to convene a local public inquiry, for example, and no legal guidelines specify the conditions under which a planning application must be called in by the secretary of state. Most important, the secretary of state is "not statutorily bound by the findings of the inquiry and is completely at liberty to overrule the hearing officer's recommendation. In doing so the Minister may be influenced by government policies that were not raised at the inquiry."[8] Nor does the system of planning appeals involve any attempt to develop a body of binding precedent: "Decisions by the Secretary are not regularly published, the view being that in land-use planning there are no 'right' answers nor rigid rules which can be applied in all cases—each case must be decided on its own merits."[9] In *North Sea Oil and Environmental Planning*, Ian Manners cites the case of an application to construct a refinery and oil storage facility at Nigg Point, on the Cromarty Firth in Scotland. After quoting the strong and vigorous objections of the reporter (as an inspector is called in Scotland) to the application, he writes:

> ... it seems unlikely (one hesitates to say inconceivable) that such a strongly worded and forceful official report (perhaps in the context of an environmental impact statement) could have been so easily overturned in the United States. Yet the Reporter's recommendations were rejected by the Scottish Secretary as not being in the national or regional interest and the plans were approved.[10]

While the courts require that local planning decisions be made according to relatively strict standards of due process, "they have been very reluctant to seem to substitute their own planning judgement for those of elected political authorities."[11] As Lord Hailsham wrote in *Chief Constable of North Wales Police* v. *Evans* (1982),

> it is important to remember in every case that the purpose of the remedies is to ensure that the individual is given fair treatment by the authority to which he has been subjected and that it is not part of that purpose to substitute the opinion of the judiciary or individual judges for that of the authority constituted by law to decide matters in question.[12]

One student of the British planning system concludes: "Broadly, therefore, the DOE, usually acting as appellate authority, is untrammelled as a planning authority so long as . . . 'procedural fairness' is maintained throughout."[13]

Parliament has played a much more prominent role in shaping British land-use and conservation policies than in controlling pollution. Some particularly controversial planning proposals, such as the selection of a site for a third international airport to serve London, have been extensively debated in Parliament, and Parliament effectively gave planning permission for both the construction of a nuclear reprocessing facility at Windscale and for coal mining in the Vale of Belvoir by laying orders before the House of Commons following the public inquiries. And owing to the personal interest of some peers in conservation issues, the House of Lords has periodically played an important role in shaping legislation affecting the countryside. More informally, planning decisions made at the ministerial level have frequently been influenced by parliamentary opinion and debate.

### Compromise

While British planning procedures have many of the characteristics of an adversarial system, in fact the Town and Country Planning Act implicitly places a high value on compromise. Writing in 1974, one commentator noted, "All recent developments in administrative procedures . . . have one common denominator—the retreat from the adversary process."[14] The Town and Country Planning Act of 1968 "adopted an informal procedure for presenting statutory plans rather than the old system of a development plan inquiry with its legalistic atmosphere."[15] Moreover, the considerable expenses and delays associated with public inquiries provide officials at both the local and national levels with a strong incentive to reach agreements that obviate the necessity for more formal or adversarial proceedings.

> . . . given the time and effort involved . . . the public inquiry procedure is not frequently invoked. Indeed, it appears to be regarded as something of a "court of last resort" to be used sparingly and only when informal discussions among the developer, the local authority, and contesting groups have failed to produce a compromise solution acceptable to all parties.[16]

Even among those planning decisions that are appealed by the developer, only a relatively small proportion become the subject of public inquiries: the rest are resolved simply through "written representations." Thus in 1981, out of 14,451 planning appeals decided by the Department of the Environment, in only 2,913 cases were public inquiries held. Another 73 were called in.[17] Since the ability of the developer to appeal his application to the minister "tends to guarantee a broad consistency in planning policies," both developers and local planning authorities are usually in a good position to anticipate the likely outcome of any particular application.[18]

Unless they have no alternative, developers generally try to avoid locations that are likely to arouse substantial community opposition. Because hearing inspectors' recommendations are often influenced by the climate of local opinion, developers tend to make a considerable effort to persuade local residents of the merits of an application before it is formally submitted. As a result, many potential planning conflicts are resolved through informal negotiations among developers, local planning authorities, and amenity groups. Planning applications are frequently modified to meet local objections before they are formally submitted.[19]

While the public inquiry process does frequently result in identifiable winners and losers, the planning process, unlike a judicial proceeding, is not necessarily a zero-sum game: particularly in cases that have aroused considerable controversy, the inspector will often recommend approval of an application, thus satisfying the developer, but insist that it be subject to a number of conditions or consents, thus attempting to respond to at least some of the objections raised by community residents and amenity groups. This is particularly likely to happen when, as is often the case, the community itself is divided. The following planning controversies illustrate this dynamic.

### Bedfordshire

In 1975, Berk Ltd., a subsidiary of the Steetley Company Ltd., applied for permission to mine for fuller's earth—a type of clay used primarily as a binding agent in foundry molds—on an 85.6-acre site approximately a mile from three Bedfordshire communities. The company required access to these deposits in order to maintain its current production, since the only other area currently being mined was due to be exhausted by the end of the decade. In a letter to the county planning office, the Department of Industry expressed its hope that "the council will be aware of the national interest when deciding the company's application." The Bedfordshire County Council replied that "the working of Fuller's Earth generally involves considerable conflict with amenity in areas of great landscape value." The conflict was relatively clear-cut: the decision to grant or deny planning permission would depend on the balance between "the need to work this rare and valuable mineral against the damage that may be done to the environment and local amenity."[20]

The Mid-Bedfordshire District Council urged that the application be refused. It based its recommendation on two factors: its doubt as to the strategic need for fuller's earth and its concern about the environmental impact of mining in the area. With respect to the latter, it concluded, on the basis of a survey of the recreational use of the area made by both local residents and visitors and of its observation of the damage inflicted on other regions where fuller's earth was mined, that the granting of the application would seriously undermine the "quietness and tranquility" of the Brickhills Recreational Area. In recommending refusal, the council described the noise and disturbance that the mining operation would cause, drew attention to the proximity of schools and convalescent homes to the proposed excavations, and noted "the unique physical features of the site."[21]

Objections to the mining operation were received from thirty-three residents, the local amenity society registered its "categorical objection," and the three local parish councils immediately affected by the proposed mining all opposed it. On February 22, 1976, the local planning committee unanimously voted to oppose the application, concluding that "it would conflict with the aim of maintaining and improving this area for public recreation and would reduce, to an unacceptable degree, its tranquility and charm." Furthermore, it "was not satisfied that the national need for Fuller's Earth outweighed the environmental damage which would be caused to this high amenity area."[22] The company appealed, and a public inquiry was held in Aspley Guise in November.

In August 1977 the inspector submitted his recommendation to the ministry. He was persuaded by the evidence demonstrating the economic importance of fuller's earth to the national economy and was convinced that alternative sites were unavailable. Moreover, he contended that loss of the area for recreational uses was not sufficient grounds to deny the application. On the other hand, he did appear to be very concerned about noise levels likely to be created by the excavation plan envisioned in the application. He thus recommended that the appeal be refused in the form in which it was submitted but added that the effect of the excavations on the local community and the landscape would be greatly lessened if the company were to reverse the direction in which it planned to pursue its operations. In his decision letter, issued in March 1978, the minister allowed the excavation to proceed but accepted the inspector's suggestion that the direction in which the deposit was worked should be altered. He also specified twenty detailed conditions that the mining company would be required to meet—conditions that were substantially similar to those proposed earlier by the county planner. As Andrew Blowers concludes in his detailed analysis of the controversy, "the mining interests were victorious on the general principle of excavation, but local opposition had secured a not insignificant victory in the actual terms of the permission . . . they were sufficiently persuasive to protect their residences from the worst possible effect of the working."[23]

Cheshire County

A similar set of tensions between amenity and economic interests was involved in the controversy over the effort of the Shellstar fertilizer works to expand its facilities in Cheshire County. Nearby residents had long complained about the noise, odors, and particulates emanating from the original plant, and they strongly opposed the granting of planning permission. A large number of people were likely to be affected by pollution from the new plant. The community residents, predominantly middle-class, organized into several active amenity groups. In response to substantial community pressure, several local districts and towns recommended that planning permission be refused. Because of the regional, even national, importance of the fertilizer works, the secretary of state called in the application. Before the formal public inquiry, the elected representatives of both the county and the district entered into negotiations with the company in efforts to establish conditions that would be applied if planning permission was granted. The company, a Dutch concern that had only recently bought the facility at Shellstar, impressed the community leaders both with its apparent ability to run the existing plant with reduced emission levels and with its willingness to accept environmental conditions proposed by county planning officials. As a result, the representatives of both Ellesmere Port and Cheshire agreed to support the company's proposal, provided that stringent safeguards were included in the conditions attached to the granting of planning permission. Christopher Wood writes: "The public inquiry held in 1976 was notable for the articulate and reasonable contribution of members of the public, in the form of local amenity societies and parish councils, and for the insistence by the planning authorities on adequate environmental controls should permission be forthcoming."[24]

The company emphasized its contribution to the British balance of payments and called attention to the local employment that its plant would offer. The inspector recommended that the application be granted, subject to the conditions agreed to between the planning authorities and the company to control noise and air pollution. He also proposed a condition "requiring the monitoring of chemical emissions from the plant." The secretary of state, however, chose to grant the planning authorities only the most minimal power to control pollution, and dismissed the conditions relating to air-pollution abatement and monitoring on the grounds that they duplicated the powers of the Alkali Inspectorate. Nonetheless, the company, in cooperation with the Alkali Inspectorate, did manage to reduce its emissions to acceptable levels, and "there have subsequently been no environmental problems from the works comparable with the pollution problems experienced in the late 1960's and early 1970's which had so incensed local residents."[25]

### Laporte

A similar outcome took place in 1974 in connection with the efforts of the Laporte Chemical Company to extend its fluorspar quarrying and processing operations in the Peak District National Park. The company requested permission to construct a new tailings lagoon to dispose of the fine particulate waste material produced by its operations. It insisted that there was no other economically feasible method of reducing water pollution and threatened to close its operations if it were denied consent. The Peak District Joint Planning Board refused permission for the lagoon, contending that the company should have developed an alternative method of disposal—one that would be likely to create less damage to what was a particularly scenic area.

The residents of the communities surrounding the facility, however, disagreed with the board's decision. District and parish councils in the area, along with the Transport and Municipal Workers unions, strongly supported Laporte because several hundred jobs were at stake; indeed, "local opinion was almost unanimously in favor of the development."[26] Arrayed in support of the board were several national amenity organizations. A public inquiry was held in 1975. The company called in expert witnesses to demonstrate that no alternative method of waste disposal was available; the Peak Park Joint Planning Board retained the services of a geologist to argue its case. The inspector recommended and the secretary of state decided that the appeal should be allowed; they concluded that there was no economically viable alternative to a tailings lagoon and that the national interest dictated the continued development of this resource. The secretary followed his inspector's recommendation, however, and required the company to meet a large number of conditions, including a time limit on the use of the lagoon for waste disposal.

## PUBLIC PARTICIPATION

While planning decisions have provided both amenity groups and community residents with some of their most important opportunities to affect British environmental policy, the concept of public participation in planning decisions is of relatively recent origin. It dates principally from the publication of the Franks Report in 1957, described as "one of the turning points in British administrative history."[27] The Town and Country Planning Act of 1968 required planning authorities to take into account the views of the public with respect to individual planning applications. A year later the Skeffington Committee, which had been formed to explore the practical implication of this aspect of the act, urged local planning authorities to work particularly closely with groups of local residents and further suggested that "the opinion of groups should be represented to the planning department through advisory committees or public meetings involving group leaders, or through the appointment of community development officers who could not only act as a link with existing groups but also stimulate formation of new groups." The acceptance of the recommendations of the Skeffington Committee signaled "the commitment of central government to the principle of participation." A subsequent government circular to local planning authorities urged "still further cooperation between local authorities and voluntary bodies."[28] The dramatic rise in the number of amenity groups over the last fifteen years can be attributed largely to the government's official endorsement of the principle of public participation in planning decisions.

The planning system's emergence as the most important vehicle for public participation in the making of British environmental policy is due both to the fact that decisions regarding land use have extremely important environmental implications and to the relatively closed nature of the British system of pollution control. Indeed, if the United States can be said to engage in land-use planning under the guise of regulating pollution, in Britain the opposite is the case: planning decisions represent the most important opportunity for members of the public to affect the level and kind of emissions to which they will be exposed. Particularly in the relatively more prosperous middle-class communities in the south of England, amenity societies and local chapters of the Council for the Protection of Rural England work extremely closely with local authorities; "increasingly planning authorities have come to rely on environmental groups in fulfilling their duties with regard to public participation."[29] Many local planning officials draw on the political and technical skills of environmental groups to assist them both in responding to planning applications and in opposing those that they consider undesirable.

In addition, the British government has officially recognized the rights of various statutory organizations to participate in planning decisions that affect their interests; indeed, they must be consulted by local planning authorities whether or not a public inquiry is held. These groups include the Countryside Commission, which is responsible for proposing land to be designated as areas of outstanding national beauty and for advising on the use of land that is part of the national park system, and the Nature Conservancy Council, which is responsible for identifying and protecting areas to be designated by the government as nature reserves and as sites of special scientific interest. Both on occasion also testify at public inquiries. Parliament has also indicated that it expects the Central Electricity Generating Board, Britain's nationalized utility, to consult with both bodies before publicly announcing any new construction plans. In addition, two national parks, the Lake District and the Peak District, have their own planning boards, which are responsible for evaluating applications for developments within their jurisdictions.

While the local public inquiry is ostensibly designed to assess the impact of a proposed development on the surrounding community, over the last fifteen years it has become increasingly used by amenity organizations as a way of mobilizing national political support for their views.

> While the prime purpose of the inquiry remains the judicial one of adjudicating between disputants, the liberalization of procedures on the standing of third parties and the admissibility of evidence has increasingly opened up inquiries to the wider role of articulating broader policy issues. Private interest groups and even public agencies have come to see the potential of the local public inquiry as a procedure for illuminating general issues through particular cases.[30]

The most important political resource available to amenity groups is publicity: the more national media attention a development proposal receives, the more likely the government is to find itself under pressure to deny approval to it. I have already discussed one such planning controversy: the construction of a third international airport for the London area. The following two cases, one drawn from the mid-1960s and the other from the early 1970s—further illustrate both the effectiveness of the pressures amenity groups have been able to generate and the relative visibility of particular planning decisions within the context of British politics.

### Rio Tinto–Zinc

In 1966 Rio Tinto–Zinc (RTZ), a multinational mining company headquartered in Britain, began to negotiate with the owners of mineral rights in the King's Forest area in the south of the Snowdonia National Park in Wales. By 1968, following extensive geological investigations, the company was persuaded that "low-grade copper mineralization of importance might exist."[31] It thereupon proceeded to engage in exploratory drilling, though without seeking consent from the local planning authority. Whether or not this consent was actually required under the Town and Country Planning Act remains in dispute; the company's action, however, created considerable resentment on the part of local officials. On April 28, 1970, RTZ formally submitted applications to engage in shallow geological drilling to four local government bodies; each application was referred to the Merioneth County Council as the local body with authority to approve all mineral developments within the county. Because the matter was of more than local importance, the county planning officer requested that the secretary of state for Wales call it in. On July 24 he agreed to do so. RTZ meanwhile continued to drill without permission, and it was not until October, after forty-eight holes had been drilled, that the Welsh Office finally ordered RTZ to cease its explorations. A public inquiry was scheduled for December.

While the inquiry was ostensibly addressed only to the issue of additional exploratory drilling, it was soon apparent that a much broader issue was at stake: What balance should be struck between the preservation of amenity interests on one hand and the development of national resources on the other? More specifically, should open cast mining on the large scale envisioned by RTZ be permitted in a national park? Underlying both these questions was a third: How should the stake of residents in additional employment opportunities be weighed against the interests of visitors to an area of unusual scenic beauty? In the months before, during, and after the inquiry, forces opposing the granting of permission for RTZ to continue its drilling became increasingly active and vocal. Spokesmen for the Council for the Protection of Rural Wales, the Friends of Snowdonia, the Ramblers Association, the West Wales Naturalists' Trust, the Nature Conservancy, and the County Council and Countryside Commission announced their opposition to RTZ's application. A national media and research campaign was launched by the Friends of the Earth around the slogan "Britain is neither rich enough to afford to sell Snowdonia nor poor enough to need to."[32] The FOE was able to secure sufficient funds to sponsor a four-man research team to undertake a highly sophisticated study of the future of mining in national parks. The controversy also received considerable national attention, with articles highly critical of mining in Snowdonia appearing in *The Sunday Times, The Observer, The Guardian*, and the *New Scientist*.

In July 1971 the Department of Trade and Industry announced a £50 million plan that provided for extensive public subsidy of mining exploration programs in the United Kingdom. Not surprisingly, within a week of this announcement the secretary of state for Wales announced the government's approval of RTZ's request to resume its program of exploratory drilling. Still left unresolved was the question as to whether the company would be given permission actually to mine copper on national park land.

Following the government's decision, the antimining campaign gathered momentum: throughout 1971 there was considerable public discussion of the future of the national parks. On April 17, 1972, Edwin Arnold, the influential mining correspondent for the Conservative *Daily Telegraph*, who had earlier strongly supported copper mining in Snowdonia, suddenly reversed his position. In an article titled "Keep the Mines Away from Our Beauty Spots," Arnold documented the immense amount of waste material RTZ's mining operation would produce and concluded:

> I no longer believe, as I once did, that mining in Britain's National Parks is such a good thing. These areas of unique beauty in our tiny crowded island must, I now feel, be preserved at all costs for the nation and posterity. ... Given the huge social costs for Snowdonia copper, I feel as a nation we can afford to forgo that much domestic mine output and buy it from the world market.[33]

The company also received another public setback: the local Transport and General Workers' Union branch voted to oppose any scheme for open cast copper mining in the area, evidently judging that the temporary economic benefits produced by copper mining were not sufficient to justify the permanent disfigurement of the local landscape.

Shortly afterward, on May 22, 1972, the BBC broadcast a film titled *Do You Dig National Parks?* which not only effectively summarized the diverse arguments against RTZ's mining plans but forced a rather reluctant RTZ into publicly defending its position. About the same time two books appeared on the subject: *Eryri, the Mountains of Longing*, a detailed case study of the Snowdonia National Park, and *River of Tears*, an indictment of RTZ's mining activities throughout the world. The former received considerable additional publicity when RTZ solicitors sought to enjoin its publication by charging that parts of the book gave "a seriously distorted picture of our client's activities and intentions in Snowdonia and are plainly defamatory."[34] The book thus received two sets of favorable reviews: once before publication and once when it was actually published. In addition, stockholders critical of RTZ's activities in both Britain and Namibia used the company's annual meeting as a forum to denounce its policies.

On April 19, 1973, RTZ announced that its preliminary geological evaluations had persuaded it to abandon its efforts to mine copper in Snowdonia. Publicly the company attributed its decision to strictly economic factors. The relatively low quality of the copper ore, it said, along with the additional environmental expenses associated with the project, rendered it "extremely doubtful that a mining operation could be economic in the foreseeable future."[35] Whatever the precise mix of political and economic factors underlying the company's decision, it was viewed as a major political triumph by amenity interests: they had prevented mining activities from taking place in a national park. Nor was the experience of RTZ lost on the rest of the mining industry. Shortly afterward seven large mining companies set up a Commission on Mining and the Environment, chaired by Lord Zuckerman, in an attempt to meet growing criticism from such groups as Friends of the Earth.

### Drumbuie

A controversy of similar political scope subsequently occurred in connection with the efforts of two British companies to construct concrete oil production platforms at Drumbuie on Loch Carron, a particularly beautiful and unspoiled deep-water loch in the Scottish highlands.

> This controversy became the cause célèbre of North Sea developments, as it pitted a small community of twenty-three people, the National Trust for Scotland and the "environmental lobby" against two large companies, the Department of Trade and Industry (and then Energy) and the Highlands and Islands Development Board.[36]

The dispute began in April 1973, when two companies, John Mowlem and Taylor Woodrow, submitted separate applications for planning permission to construct concrete production platforms at Par Cam, Drumbuie, on land owned by the National Trust of Scotland. Their application contended that "Drumbuie was the most suitable location on the stated coastline, possessing a necessary combination of a sheltered bay, deep water, and adequate level ground onshore."[37] Drumbuie, however, was a small, isolated hamlet whose handful of inhabitants were

> steeped in the crofting way of life and in old-fashioned traditions of strict religious observance and family and community ties.... Opponents of the ... proposal expressed fear that an influx of cosmopolitan workers and sophisticated businessmen demanding Sunday work... would upset their traditional ways.[38]

121

Complicating the issue was the fact that the land in question had been left "inalienably" to the National Trust for Scotland, and thus an act of Parliament would be required before development could proceed. Given the large number of objections to the proposal, the application was immediately called in by the secretary of state for Scotland and a public inquiry was scheduled.

The inquiry lasted from November 1973 to May 1974. Testifying in support of the application were the two construction companies, the Department of Trade and Industry, and the Highland and Islands Development Board, the official government body whose purpose is to promote development in Scotland. They argued that no other site met the requirements for the construction of the kind of platform the oil industry needed to develop North Sea oil. Representatives of the British government expressed particular concern that unless this construction site was promptly approved, orders for the construction of the more than forty concrete platforms estimated to be required for the development of the British sector of the North Sea would be placed in Norway instead of Britain. The government estimated the total revenue likely to be produced by these orders at £800 million over the next decade.

The primary objectors at the hearing included the County Council for Ross and Cromarty and an ad hoc organization of the people who lived and worked in the immediate area of the proposed project known as the South West Ross Action Group. Three national environmental organizations—the National Trust, the Conservation Society, and the Friends of the Earth—also testified in opposition to the application. Some of the heated tone of the proceedings—which were estimated to have cost all parties £350,000—is suggested by the following exchange between the applicants and the local residents:

Q. Is it a fact that the land you want is at present held by the National Trust for Scotland?

Yes.

Q. Is it a fact that it was bequeathed to the Trust by the late Lady Hamilton?

(pause) Yes.

Q. Is it also a fact that you have no scruples about setting at naught a dead person's last will and testament?

During the ensuing sudden consultations among the industrialists, the hall erupted in applause, closing ranks behind the questioner. The affirmative reply, when it eventually came, was greeted with dismay by people who still attend church regularly on Sundays and hold fast to the old Biblical irtues.[39]

In the midst of the public inquiry, the focus suddenly shifted to London. The Conservative government announced its intention to introduce a bill in the House of Commons that would permit the government to nationalize all sites considered to be urgently needed for oil development, thus entirely circumventing the planning process. While the bill did not refer to any particular development, it was obvious that it was aimed primarily at Drumbuie. In fact, it immediately became known as the "Drumbuie legislation" and attracted considerable editorial comment—both favorable and critical—in the British press. Introducing the bill, the secretary of state for Scotland informed Parliament:

> In the exceptional circumstances of our national need to obtain as much oil from the North Sea as we can as soon as possible, we have decided to introduce a bill which would enable the Government to acquire, using an accelerated procedure if necessary, land which is urgently needed for certain projects related to the production of offshore oil and gas. The Government would then lease the land to operators with appropriate safeguards as to its use. The accelerated procedures would apply also to planning permissions required for the projects themselves or essential and urgent supporting activities.[40]

Behind the government's decision to introduce the bill was the pressure it was under from both the OPEC oil price increases and the strike of British coal miners; indeed, the initiative to introduce the legislation came from the newly created Ministry of Energy, headed by Lord Carrington. The bill met with little enthusiasm in Parliament, however. It particularly outraged Scottish members, who were upset by what they regarded as the usurpation of traditional Scottish planning powers. One Labour member from Scotland inquired:

> Is the Secretary of State aware that there is considerable dissatisfaction in Scotland about the terrific hurry to exploit the oil, which has been in the North Sea around Scotland for millions of years? Does he realize that such speedy exploitation of the oil as he seems to envisage could do irreparable damage to Scotland, not only to Scotland's beauty but to its economy, through our not knowing the full facts about how the oil should be exploited, which is a very difficult technological task?[41]

The legislation became moot, however, when a Labour government was returned to power in early 1974.

The reporter submitted his recommendation to the new Labour Scottish secretary, William Ross, and on August 12 Ross concluded that permission should not be granted. He stated:

This balance of judgment must be made in each case, but my general approach is to look favorably on applications for technically suitable sites which have access to existing facilities, which can draw on existing sources of labour, and which make use of existing infrastructure and services. Correspondingly, difficulties must be expected over applications for sites which lack these.[42]

This decision was widely applauded in Drumbuie, whose residents spent the next evening passing the bottle from house to house to toast the name of "Willie" Ross and celebrate their David-and-Goliath victory.

## Implications

These conflicts over land use were unusual not only in the degree of controversy they created but also in their outcomes. As we have seen, the more usual result of planning controversies in Britain is some sort of compromise. Development is usually allowed to proceed but is subject to constraints designed to meet at least some of the objections of amenity groups and community residents. In the cases of RTZ and Drumbuie, a unique factor was responsible in each instance for the victory of antidevelopment forces. In the case of RTZ it was the fact that development permission was being planned for an area designated as a national park. While development is not forbidden in national parks and 98 percent of the land within them is privately owned, the parks do enjoy a unique status as part of Britain's "community property." The national park

system has kept at bay changes which might have destroyed the sense of remoteness and peace and quiet for which people value the areas the national parks cover. The successful opposition to plans to put a motorway through the heart of the Peak Park in 1977, *to mine copper in Snowdonia in 1972*, to extract potash from the North York Moors in 1979 and to afforest a large area of moorland in Exmoor in 1957 all relied heavily on the national park status of the threatened countryside.[43]

Significantly, more than half of the British government's spending on conservation and recreation in the postwar period has gone into the national parks. Similar presumptions against development are also enjoyed by land designated as nature reserves by the Nature Conservancy Council, land designated by the Countryside Commission as areas of outstanding beauty, and in the "greenbelts" surrounding many of Britain's urban areas in the structure plans of local planning authorities.

In the case of Drumbuie, the fact that the land in question was owned by the National Trust was extremely significant. The land owned by the

Trust constitutes Britain's "core" countryside and encompasses many of its most scenic areas, particularly on the coasts. It also enjoys more legal protection from development than any other land in Britain; only once in the last seventy years has Parliament reversed the inalienability of land given to the Trust. In fact, immediately following the Drumbuie inquiry, an oil production platform was approved for Loch Kishorn, even though there was no reason to distinguish between the two sites on amenity grounds. The Loch Kishorn site, however, was not owned by the National Trust of Scotland. (There is considerable overlap between national park land and land owned by the Trust; more than two-thirds of the former is owned by the latter.)

## THE POLITICS OF PLANNING

### The Advantages of Developers

While the Town and Country Planning system has allowed amenity groups considerable opportunity to participate in the making of public policy, the position of developers nonetheless remains a privileged one. The most obvious advantage enjoyed by developers is that they alone have the right to appeal local authority decisions. Amenity interests cannot even legally demand a public inquiry. If a local planning authority favors a particular application, the only way it can be challenged is for the secretary of state to call it in. And the secretary is unlikely to do so unless the objectors have managed to generate so much opposition to it that the case has acquired the status of a controversy.

> Almost all controversial planning proposals are the subject of a local planning inquiry. There must, however, be evidence of controversy. Thus there was no public inquiry over the application to build a nuclear power station at Heysham, near Lancaster, despite the fact that it has been the only nuclear power station to be built close to a large population area. In this case there were few objections at the time of the proposal.

Moreover, given the limited resources of Britain's amenity organizations,

> some planning proposals may go unnoticed while the energy of potential opposition is directed elsewhere. The oil production platform application at Loch Kishorn is an example. This application was approved without a public inquiry although the inquiry inspector at the Drumbuie public inquiry considered its beauty to surpass that of Drumbuie itself.[44]

Even if objectors are sufficiently vocal to secure a public inquiry, either because they have persuaded the local planning authority to deny an application or because it has been called in, the developer still enjoys a considerable number of advantages. A critical one consists of knowledge. "Only [the developer] knows why he needs the site, and what the economics of its exploitation are, for he is under no obligation to make known to the public or even to the local planning authority anything more than the fact that he wants to develop a particular site for a simply defined purpose." If a development is to be successfully opposed, challengers must convincingly demonstrate either that it is altogether unnecessary or that some alternative site is actually preferable. To be successful, either of these assertions usually must be supported by the testimony of professional engineering and economic consultants. "A developer may allege that no alternative site could possibly suit him, for this or that reason. If the objectors cannot produce witnesses who can convincingly refute his claim, the Inspector will tend to accept the developer's word."[45] The hiring of such consultants, however, frequently costs more than community residents or amenity groups can pay. As Anthony Barker concludes,

> in essence the concept of what might be called "procedural fairness" (offering a stage in the inquiry for these parties and interests to have their say) holds sway rather than any idea of "substantive or real fairness" which would ensure that all those who may legitimately be heard at the inquiry actually have the means to make their questions and assertions effective through access to expert advice or information which often, of course, requires some access to funds.[46]

The developer enjoys one other important advantage over amenity interests: he can often count on considerable support from those residents and their families who stand to benefit from the employment opportunities provided by investment in new facilities or the expansion of existing ones. In view of the substantial unemployment prevailing in parts of Great Britain over the last two decades, the prospect of new jobs has assumed considerable significance. The likelihood that controversial developments will secure planning approval is substantially greater when they are proposed for a relatively economically depressed region or community: either the local planning authority will be swayed by community or trade union pressures to overcome the objections of amenity groups and grant initial approval or the minister will find it easier to justify planning approval on the grounds that he is responding to local community sentiment.

As a result, public access to the decision-making process by no means guarantees that environmental interests or amenity values will triumph. In the case of pollution control, the lack of opportunities for public participation may mean that less weight is given to environmental considerations; in the case of the planning system, however, it is often precisely the relative accountability of local planning authorities to public pressures that undermines the political influence of amenity interests.

### Bedfordshire Brick

A particularly dramatic and unusually well-documented case illustrating this dynamic occurred in connection with a proposal to replace the brickworks in Marston Vale, Bedfordshire, with a new facility in the late 1970s. Approximately 20 percent of all the bricks manufactured in Great Britain are produced at the London Brick Company's facilities in Bedfordshire. The making of bricks from Oxford clay produces three major pollutants: sulfur dioxide, fluorides, and mercaptans. The first two are health hazards; the latter emits a foul odor. Although public concern about these emissions surfaced periodically during the 1950s and 1960s, the externalities of brick production tended to be overshadowed by the overriding national need for cheap bricks to increase the nation's housing stock. However, public complaints about air pollution increased significantly in the early 1970s, and in fact Bedfordshire acquired a reputation as "one of the most polluted parts of rural England."[47]

The Alkali Inspectorate continued to insist that London Brick was employing the best practicable means of minimizing the emissions to which local residents were exposed. (Since the technology did not exist to reduce the emissions themselves, the Inspectorate had required them to be dispersed into the atmosphere through the construction of tall stacks.) When complaints persisted, particularly from local farmers upset about the effect of fluoride compounds on the health of their cattle, the parliamentary under secretary of state at the Department of the Environment, Eldon Griffiths, established a liaison committee to examine ways of improving control of air pollution. It consisted of the Alkali inspector, local authorities, and representatives of industry. It did not, however, appear to produce any noticeable improvement in the behavior of the brickworks.

In 1978 the London Brick Company (LBC) announced its intention of radically transforming its facilities. It proposed to replace its two main existing brickworks with two new works, each equipped with chimneys 400 feet tall. While the same amount of effluent would be emitted, it would now be diluted and spread over a wider area. The company further announced that it was prepared to negotiate with the County

Council to accept new conditions on its operations. Anticipating rapid approval of its plans, in August 1979 the company formally applied for planning permission. The managing director subsequently recalled, "We thought the plans would be snatched from our hands and we'd be told to get on with it as soon as possible."[48]

What the company failed to recognize was that public resentment over the pollution problems caused by the brickworks had been steadily accumulating for more than a quarter of a century. Now that local residents finally had an opportunity to register their disapproval of LBC's pollution-control efforts, they were determined to seize it. As an officer of the country planning authorities observed early in 1979, "the public debate that has taken place in recent weeks has revealed widespread concern about the possible pollution effects of the LBC's operations." A few months later he noted, "It has become increasingly apparent that the polluting emissions from the brickworks and their impact on human health, agriculture and wildlife is the prime issue."[49]

A report by environmental consultants provides some ammunition for both sides. While it reported that "there is little likelihood that the emissions from the existing works are significant with respect to human health" and that their effect on plants and animals was difficult to assess, it also suggested that "the loss of amenity and psychological discomfort due to the brickworks' odor is likely to be of most significance." The consultants concluded that the application should be deferred pending investigation of the feasibility of new techniques for removing the odor. LBC replied that its research staff had already attempted and failed to develop a more satisfactory abatement technique and that further research would serve only to delay development plans and result in "a crippling effect on the national market for Fletton bricks."[50]

By the time this report was released, attitudes had begun to harden. The opposition to county approval of the company's plans was led by a newly organized group called PROBE (Public Review of Brickmaking and the Environment). Chaired by the Marquis of Tavistock, a prominent local landowner, the group included three local MPs, three members of the Bedfordshire County Council, and representatives of academic, medical, farming, and business interests. Its secretary was a local farmer whose cattle had contracted fluorosis. PROBE was subsequently supported by the National Farmers Union, whose members had become increasingly antagonistic toward the brickworks. Supporting London Brick's proposal were the County Council's professional planners, the chairman of the Environmental Services Committee, the Alkali inspector, and official medical and research opinion. The Inspectorate argued that the environmental risks were minimal and would be substantially reduced by the construction of tall stacks. It contended further that the new facilities would offer both "effective controls and long-term security for an industry which contributed to local wealth and employment."[51]

In July the full council met to consider the company's application. Following a prolonged, heated, and at times rather eloquent debate, the council voted 38 to 28, with 17 members either absent or abstaining, to grant planning permission on the condition that the brick kilns were "designed to be capable of removing the pollutants and odours given off in the firing process." The company responded to the council's stance in a letter urging the council to interpret the phrase "capable" in terms of the principle of the best practicable means. Since the council had now approved its application, LBC argued that its planning condition should "be both reasonable and not repugnant to the permission." The chairman of the London Brick Company personally reiterated the company's condition at a subsequent meeting with members of the council. He argued that LBC's proposals would bring immediate environmental benefits and offered the possibility of further improvement in emissions if and when improved pollution-control technology was developed. Somewhat more ominously, he implied that the company might not appeal against the condition, stating that "it was a matter for the people of Bedfordshire to decide." He then added: "We have many other investment opportunities, we have existing permissions for new works outside the County and I am not saying that in any way as a threat."[52]

The full council then reconsidered the entire matter at its meeting in December. There appeared to be considerable resentment at LBC's attempt to undermine the council's previous decision; one member characterized the company's tactics as "a disgrace" and "an insult to the Council."[53] By a vote of 43 to 33 the council reaffirmed the decision it had reached six months earlier. Those who wanted to use the planning process as a way of imposing stricter pollution-control standards on the brickworks than those demanded by the Alkali Inspectorate appeared to have triumphed.

This victory, however, proved to be a Pyrrhic one. Immediately after the second vote the company announced that it would not undertake any new investments in Bedfordshire, but would instead invest in two communities where it had secured acceptable planning permissions. Two months later LBC announced that it was closing the Ridgemont works, terminating 1,111 people. Although the company cited economic reasons for the move, it was widely believed that the decision to close the works at Ridgemont was "a deliberate attempt to associate the closure with the Council's refusal to grant an acceptable permission." With unemployment suddenly replacing pollution as the critical issue, local trade union members began to pressure the council to reverse its decision. After a meeting among representatives of the council, the company, and the unions, a compromise was reached. The Environmental Services Committee agreed to grant approval for the construction of a new plant at Ridgemont without conditions regarding pollution control while the company promised that Ridgemont would become the first priority for redevelopment. Andrew Blowers, who has described this conflict in considerable detail, concludes: "These assurances, anxiety about jobs, and the fear of reprisals at the coming elections in May concentrated the minds of councillors and permission was granted for the replacement works ... without referring the matter to the Council."[54]

*The Role of the Central Government*

The ability of amenity interests to use the planning system to impose strict environmental controls on the development of industrial and natural resources suffers from an additional obstacle: the role of the central government. British planning law gives developers the right to appeal the denial of their applications by a local planning authority to one of Britain's three secretaries of state. The latter can also call in planning applications that are deemed to be of national importance, thus bypassing the local planning authority. The purpose of providing ministers with this authority is only partially to ensure that the criteria used in deciding on planning applications are relatively uniform throughout Great Britain; in fact the system continues to exhibit considerable regional variation, both among and within England, Scotland, and Wales. The far more important reason is to ensure that the interests of the nation as a whole will take precedence over the preferences of particular communities.

While community interest may or may not include amenity considerations, the national interest is almost invariably defined in economic terms—often specifically in terms of the balance of payments. Developments that are strongly supported by local residents will almost never be denied planning approval by the central government, as they often are in the United States. The opposite, however, is not the case: on a variety of occasions the opposition of local community groups to particular proposals has been overruled by Whitehall in the interests of the national economy, even though various conditions have often been attached to the applications in an attempt to respond to at least some local objections. The following case—also associated with energy development in Scotland—illustrates this dynamic.

In early 1977 Shell UK Esso and Esso Chemical Ltd. officially applied for permission to construct several related facilities for the processing of natural gas from the North Sea: a liquefied natural gas (LNG) separation plant, an ethylene cracker, and assorted pipeline and berthing facilities. The site selected by the oil companies was among those previously identified by the Scottish Development Department and several local authorities as appropriate for large-scale industrial development. It was particularly suited to the needs of the petrochemical industry as it was located only four miles from an eminently suitable berthing location at Braefoot which could be used for the export of liquid cargo. Because of the national importance of the project—it represented the first major new downstream processing plant to be constructed in connection with the North Sea—it was, like most major new oil-related applications, immediately called in by the secretary of state for Scotland.

Local residents were divided. The principal source of opposition to the application came from Aberdour and Dalgety, two middle-class communities located approximately a mile from the proposed LNG facility. The Aberdour Residents Association and the Dalgety Bay Ratepayers Association combined to form an action group that led the opposition to the oil companies' proposal. While their opposition was originally based on a combination of environmental amenity and safety grounds, the latter rapidly became dominant. They argued that the risks associated with the operation of an LNG facility were too great to permit it to be located so close to a residential area. Whether the members of the action group genuinely believed that the risks associated with an LNG facility were unacceptably high or whether "safety was being used strategically as the most likely means of ridding what was undoubtedly a quiet and attractive stretch of Forth coastline of industrial development" is difficult to judge. While individual motivations certainly varied, it seems reasonable to conclude that "the Action Group were fighting against any plant at Braefoot Bay rather than fighting for an acceptably safe plant. . . . [For them] an acceptable safe installation at Braefoot Bay and no installation at Braefoot Bay are one and the same things."[55]

The local district and regional councils, which represented primarily working-class communities with substantial unemployment, strongly supported the Mossmorran–Braefoot Bay development. They hired a firm of chemical, engineering, and scientific consultants, Cremer & Warner, to advise them on the safety and environmental nuisance aspects of the planning applications; its findings were cited extensively in the hearings. The final actor involved in the controversy was the Health and Safety Executive (HSE), which is responsible for regulating the safety of industrial establishments in Great Britain. Cremer & Warner issued a report that concluded "that if each plant was designed, constructed, installed, operated and maintained to the highest standards currently available in the industry, no intolerable situation should be imposed within the site or in the surrounding neighborhood."[56] This position was echoed by the HSE. The action group, however, strongly challenged the conclusions of both Cremer & Warner and the HSE. They argued that the latter's assessment of the possible risks to both the public and marine life as a result of a spill were far too modest and that no attention had been paid to the possibility of a collision among the vessels anchored off the terminal.

The reporter was not persuaded by the arguments of the action group and recommended that the proposal be approved. He argued that

> the plants can be designed and operated so that they should not result in an undue hazard effect on the community.... The acceptable degree of risk which a community should be asked to accept should be such that a dangerous incident which would cause injury to a member of the public outside the site boundary should not occur more than once in one million years.... Doubts which were raised on safety factors are not sufficient to cause me to believe that these plants cannot be designed to operate within an acceptable level of hazard.[57]

The secretary of state for Scotland accepted the reporter's advice and gave his provisional approval in March 1978. His decision letter stated that

> there can be no question of economic need for the developments being balanced against this [hazard] factor: considerations of public safety would automatically rule out the developments if it were shown that they would give rise to an unacceptable level of hazard. The Secretary of State is satisfied that the plants can be designed to operate within an acceptable level of hazard.[58]

Forty-eight conditions, however, were attached to the approval of each of the two applications. The most significant was that "a full hazard and operability audit should be undertaken to the satisfaction of the Secretary of State before the LNG plant and ethylene cracker could each be commissioned."[59]

The secretary of state's decision was provisional rather than final because a new safety issue had unexpectedly arisen after the conclusion of the public inquiry. The action group cited evidence that radio sparks might produce an explosion of the liquid natural gas. This matter was then investigated by both Cremer & Warner and the HSE. In August 1979 the secretary of state announced that he had received sufficient information to enable him to make a decision. The hazard of radio sparks, he said, was not significant enough to cause him to reopen the inquiry, furthermore, he had determined that the various safety audits referred to in his provisional decision had now been completed to his satisfaction. Accordingly, all applications were formally approved.

On balance amenity interests suffer from an important handicap in attempts to influence those government departments that are most likely to favor the granting of planning approval. These departments, such as the Department of Industry and the Department of Trade, represent important sources of intragovernment pressures on the decisions of the various secretaries of state. Moreover, both the Department of Environment and the Scottish Office have administrative units responsible for promoting economic development; the latter's office is actually responsible for the economic development of Scotland, while the former was, until the establishment of a separate Department of Transport, also in charge of constructing motorways (highways). The Countryside Commission and other governmental bodies with which environmental groups enjoy influence have only an advisory role in the planning system. Moreover, following the conclusion of a public inquiry, no further lobbying or debate is permitted; indeed, the inspector's report is usually not published until after the minister's decision is announced, thus providing the public with no opportunity to comment on it. As a result,

> the public inquiry structure ensures that, while an issue may be opened up to wider participation, analysis of that debate is kept strictly confidential, and the tidy control and termination of participation is ensured. It is political control with the benefits of judicial authority. Furthermore, under British constitutional principles, it is virtually immune from review in the courts.[60]

Moreover, since the mid-1970s the British central government has attempted to reduce the ability of local planning authorities to prevent developments of which they disapprove. In 1974, frustrated by the delays produced by the processing of various planning applications in connection with the development of North Sea oil, the newly elected Labour government proposed legislation limiting public inquiries at which "energy policy could be jeopardized by time-consuming attention to environmental and local interests." This proposal appeared to reflect a "diminishing political commitment to environmental protection in the face of economic difficulties."[61] The Offshore Petroleum (Scotland) Act, enacted by the Labour government in 1975, granted the secretary of state for Scotland the power to appropriate any lands that might be required to facilitate the development of North Sea oil. The Open Cast Coal Act of 1975 and the Community Land Act of 1976 also limited the right to public inquiry. And in 1975 the secretary of state for the environment rejected the recommendations of the Dobry Report, which had urged a number of changes in the planning system designed to increase public participation. (Among its suggestions were greater publicizing of development applications and allowing inspectors to recommend that costs be awarded against developers whose applications had failed, so that individuals and community groups might be reimbursed.)

The government also altered its policy toward mineral exploration in the national parks. The secretary of state overruled the decision of the Park Board to deny Imperial Chemical Industries permission to extend its Tunstead limestone quarry in the Peak District National Park, and in 1980 he awarded RTZ a license to explore for minerals in the Peak District. These decisions suggest that the government's "policy is to allow market forces and considerations to be the main generators of the programmes for extracting minerals."[62] They thus represent an important reversal of the "Silkin standard" (see Chapter 1), which contained an explicit presumption against mineral working in the national parks. In the early 1980s the Thatcher government issued a series of circulars that encouraged local authorities to expedite the processing of those planning applications likely to make the greatest contributions to the local and national economy. Both of these policies were strongly criticized by environmental organizations.

Nonetheless, amenity interests continue to exercise considerable influence over land-use policies. The Thatcher government was forced to retreat from its effort to increase the amount of land open for new housing construction by allowing building to take place on some of the greenbelt areas surrounding major urban centers. Its greenbelt draft circular was vigorously and successfully opposed by local councils and conservation groups that regard greenbelt designation as their most important weapon against suburban overdevelopment and by property-owning environmentalists eager to "pull up the ladder behind them." The *Economist* wrote:

> Planning ministers who attack "green belt" risk being classed with dog-haters and child batterers. The rings of protected countryside around Britain's major cities, first designated in 1938, are one of the few successes of modern British planning. Despite occasional inroads into them, they are a hiatus of greenery between solid suburbia and sprawling satellites of surrounding urban land.[63]

Nevertheless, pressure from developers to increase the amount of land available for housing can be expected to increase, and some modification of the government's land-use policies is likely.

## PRESSURE FOR CHANGE

### *Motorway Construction*

Ironically, because the planning system, unlike the system of pollution control, includes an element of popular participation, it has been the focus of considerably more criticism by British amenity organizations: they have sought—not without some success—to make it more accountable.[64] The issue of public participation that has prompted the most controversy has involved the procedures governing the approval of new motorways. Beginning in the early 1970s, opposition to new motorway construction increased dramatically in many communities in England. Objectors challenged not simply particular projects but the ways in which decisions about them were being made. They argued that the public inquiry procedures followed by the Department of the Environment and the Department of Transport were unfair because participants were confined to debating the merits of alternative routes; what they could not do was to question the assumptions on which the government's road proposals were based. At a local inquiry into the Denton relief road (M67) in 1975, for example, "the DOE's Inspector was asked to disregard any evidence submitted in relation to the need for the motorway."[65]

In the minds of many objectors, this limitation severely undermined the legitimacy of the planning procedures by which new motorways were approved. Opponents of British motorway policy contended that national policy matters should not be excluded because they really had never been adequately formulated and debated; what passed for transport policy was simply the assumptions, forecasts, and projections of departmental civil servants. Moreover, much had changed since the 1970 release of the white paper *Roads for the Future*: "There had been the energy crisis of 1973 and a widespread public reaction against motorways which its critics felt ought to be considered in national policy." In addition, they strongly resented their inability to challenge the accuracy of forecasting techniques and cost-benefit analysis on which both national and local policies were based. Lord Foot, the chairman of the Dartmoor Preservation Association, told the House of Lords in 1976 that at a local inquiry in Plymouth in 1974 we were presented with a multiplicity of forecasts and extrapolations as to what would be the traffic-flows in the years 1980 and 2000 coming from all kinds of different sources. Plymouth City Council had their own extrapolation which differed fundamentally from those of the Department of the Environment—and, indeed, one witness from the Department of the Environment was in favour of a different one. It is absolutely intolerable that objectors to a scheme who say, 'there is no need for this' should not be allowed to challenge the assumptions which the Department of the Environment choose to make'.[66]

They also objected to the fact that the inspectors at local inquiries into new road construction were appointed by the same department that had called the inquiry: the Department of the Environment was thus both judge and jury. Finally, they complained that those who objected on amenity grounds had no legal right to be heard; they would only be listened to at the discretion of the inspector.

Throughout the first half of the 1970s, objectors to new motorway proposals found themselves increasingly frustrated by their inability to challenge the government's road transport policies. The fairness of the public inquiry procedures was unsuccessfully challenged by a number of lawsuits. In a popular book titled *Motorways versus Democracy*, John Tyme, a member of the National Transportation Working Party of the Conservation Society, which had played a prominent part in organizing local opposition to various motorway schemes, wrote, "Democracy is endangered by the overwhelming power of the roads imperative and its successful subversion" of Parliament's powers. "The corruption of the power of decision-making in transport has led to a dangerous erosion of respect for the rule of law amongst large numbers of people; this is no longer a matter for dispute and should be cause for growing alarm."[67] In a revealing indication of the depth of the frustration felt by many community residents, local public inquiries in Eastleigh, which is near Sussex, and in London were physically disrupted. Local protests over particular motorway schemes were supported by national environmental organizations and in 1974 thirty groups opposed to additional motorway proposals joined to form the National Motorways Action Committee.

In December 1976 the government responded by appointing a committee to inquire into the justification for motorways and truck roads and the official traffic forecasts on which they are based. In April 1978 the Department of Transport and the Department of the Environment, after consultation with the Council on Tribunals, issued a white paper that went a long way to satisfying the demands for greater public participation and apparent fairness. Objectors will have access to the same information for considering future road proposals as the Department of Transport, and this will be available well in advance of the public inquiry. Moreover, alternative routes will receive much greater attention.[68]

While the government remained unwilling to allow the public inquiry system to become a forum for debating national transportation policy or to subsidize the costs of successful objectors at public inquiries, "these recommendations went a long way toward increasing the level of information on which policy decisions were to be debated" and thus constituted an important victory for objectors to Britain's motorway construction program. The advisory committee's report was regarded by such groups as Friends of the Earth as vindicating the protests of motorway objectors and exposing "the heavy-handed methods of the Department of Transport road planners."[69]

*The Large Public Inquiry*

The controversy over additional motorway construction has diminished in recent years, in part because of budgetary constraints: the Department of Transport has fewer funds available for new construction projects. At the same time, the controversy over the fairness of public inquiries has, if anything, increased in intensity. It has, however, tended to focus less on the procedures for deciding on planning applications submitted by the private sector than on those initiated by the central government or its various statutory authorities. These applications have usually involved such projects as motorways, airports, water reservoirs, new coalfields, and power stations, nuclear as well as nonnuclear. The controversy created by these developments has added a new dimension to the British planning system: the "large public inquiry." These inquiries are distinctive in both their national visibility and the degree of controversy they have aroused: they have become a forum for some of the most bitter and prolonged disputes over British environmental policy.

While the origins of the large-scale public inquiry can be traced to the extrajudicial inquiry conducted by the Roskill Commission between 1968 and 1971 (see Chapter 1), the first formal large public local inquiry took place in 1977 in connection with a controversial scheme by the British nuclear industry to undertake the thermal oxide reprocessing of nuclear wastes at Windscale.[70] That it took place at all represented an important victory for antinuclear forces in Britain, as the government had been reluctant to convene an inquiry in the first place. Since then there have been major public inquiries into the National Coal Board's request to engage in mining in Selby and around the Vale of Belvoir, the British Airports Authority's proposal to construct a fourth terminal at Heathrow, the development of an airport at Stansted (discussed in Chapter 1) and the extension of the nuclear power facility at Sizewell. Virtually all those inquiries have involved development sponsored by state or quasistate authorities. These major inquiries bear a close resemblance to American rule-making procedures: they are prolonged, highly contentious, and governed by well-developed rules of procedure.

> Big public inquiries often run on for weeks or months, sometimes become highly acrimonious, and absorb enormous quantities of official time, energy and resources. Barristers for the various interests must be present for the entire time. In these respects, inquiries go far beyond informal American rule-making procedure where at most only oral arguments are allowed, not direct and cross-examination of witnesses. Finally, the inspector recommends a decision to the Minister who must state reasons for his ultimate decision, much like the requirement of reasoned decision for rules in America. Following the Minister's decision, the inspector's report is disclosed and the order is judicially reviewable for a limited time.[71]

Aside from the obvious conflict of interest between the government's roles as judge and plaintiff, the fairness of many of these large public inquiries has been criticized on a number of grounds. One is the unwillingness of the government to help defray the costs of those who object to the programs of its agencies. Since many of these proposals are highly technical, generating the data necessary to challenge the government's recommendations often is very expensive. And while the resources of objector groups are limited, those of the government and its various agencies are not. Moreover, the government also has at its disposal the technical expertise of the entire public sector. At the Windscale inquiry, for example, British Nuclear Fuels Ltd., a state-owned company, spent approximately £750,000 to prepare its case; those who objected to the construction of the nuclear reactor had barely one-tenth that amount at their disposal.

British law and practice treat objectors to private- and public-sector development schemes in an identical manner; neither group is entitled to any financial assistance from the government, even though in the latter case the public is, in effect, subsidizing the petitioners as well. (The only exception is the procedure governing private bills, in which a share of the legal costs of the petitioners is paid out of government funds.)

Anthony Barker notes that while most local public inquiries are regarded as quite fair,

> the major problem, and even crisis, of the inquiry system arises . . . at the top end of the range. Can the local public inquiry cope with the vast breadth and burden of material of the very biggest cases? Can the moral element of fairness, impartiality, etc. be maintained while the British government refuses public funding for any objectors or other voluntary parties who wish to criticize its own direct developments or those of its nationalized or statutory agencies?[72]

Aside from the issue of intervenor funding, environmental groups have criticized the large public inquiries on the grounds that the government has not made available sufficient information to enable the public to assess the merits of alternative policies. They have also contended that their testimony on particular projects—particularly those involving nuclear energy—have not been taken seriously by the officials responsible for shaping Britain's energy policies and that the inquiry process has not been sufficiently integrated into parliamentary decision making. David Pearce concludes that

> public inquiries can no longer be judged in terms of the efficiency with which they 'advise and inform' a Minister of some authority. They must be judged on the much wider basis of the extent to which they contribute to the democratic process. For that contribution to be significant it becomes necessary to link the planning system to Parliament in a formal manner; to charge public inquiries with the right, which may not be exercised, to consider 'need'; to find ways of embracing a much wider public than the professional elites who tend to dominate the current inquiries and to provide a focus for information on the technological change that is causing the concern over the adequacy of the system.[73]

The controversies raised by these inquiries will doubtless continue.

### Agricultural Controls

During the late 1970s the British planning system came under criticism for another reason: its lack of control over agricultural land use. The general development order issued under the Town and Country Planning Act of 1947 defined "development" in such a way as to exclude changes in the countryside from its purview. Accordingly farmers ordinarily do not require permission to destroy hedgerows, cut down trees, drain wetlands, and plow over roughlands, downs, and moors. The plan-

ning system excluded agriculture because both planners and conservationists regarded agricultural development less as a threat to amenity
values than as a land use that itself needed protection from developers.
Indeed, a member of the Council for the Preservation of Rural England
described agriculture as "the least changing of industrial pursuits."[74]
During the 1930s and 1940s the CPRE concentrated its efforts on protecting the countryside not from farmers but from ribbon housing development, mineral excavation, the building of factories, offices, and
seaside bungalows, and the erection of roadside advertising. Moreover,
even if agriculture had been regarded as a threat to the beauty of the
countryside, at the time the planning system was established Britain's
wartime experience would have discouraged the enactment of any restrictions on domestic food production.

Local planning authorities do have some limited controls over agricultural land use. They can, for example, issue tree-preservation orders
if in their judgment "the felling of trees and woods ... would have a
significant impact on the environment and its enjoyment by the public."[75]
The procedures for doing so are extremely complex, however, and if
such an order is issued, the local planning authority is required to pay
the landowner compensation for lost income. In practice, tree-preservation orders are applied almost exclusively to tree removal in urban
areas.

In spite of bitter protests from the National Farmers Union, the Countryside Act of 1968 did empower the secretary of state for the environment to impose an order over moorlands located in national parks;
farmers were required to give six months' notice before either plowing
up or improving rough moorland. If a national park authority decides
that such a change is not in the interest of the surrounding community,
it has three options: it can attempt to persuade the Ministry of Agriculture to deny funds for "upgrading" the land, it can enter into a special
agreement with the landowner requiring him not to improve the land
in return for compensation, or it can persuade the owner to sell the land
to it. The first option is virtually never successful, and while the latter
two have frequently been exercised, they are extremely expensive and
thus can protect only a limited amount of acreage.

The Nature Conservancy Council, the government agency responsible
for safeguarding the interests of science in land use, does have the right
under the Countryside Act of 1968 to make agreements with landowners
to get them to refrain from carrying out various changes on land designated as sites of special scientific interest. But it has no power of compulsion and must pay farmers appropriate compensation. Finally, the
Countryside Commission, the governmental body specifically responsible

for the status of the British countryside, has no statutory authority to control land use and must rely on its powers of persuasion to urge farmers to be more sensitive to amenity considerations.

Consequently, the more than 70 percent of the land area of England that is farmed is without any effective controls over its use. And because British government policy in the postwar period has encouraged maximum agricultural output—in part by giving farmers generous subsidies—the landscape of much of England and Wales has become transformed. Perhaps the most visible sign of the commercial development of the countryside has been the steady destruction of hedgerows. In a popular book published in 1980, *The Theft of the Countryside*, Marion Shoard writes:

> Hedgerows provide the framework of the English countryside [and] have defined field and territorial boundaries in our countryside from Saxon times. And the pattern they have imposed on our fields, built up gradually over thousands of years, does more to distinguish our landscape than any other feature. By providing abundant cover, hedgerows have made England's wildlife richer by far than that of other lands: the primrose and the violet, the hedgehog and the dormouse owe their abundance to the hedgerow; and it is our network of bushy hedges which has made England *par excellence* the country of small songbirds. Most important of all, however, it is the hedgerow that has given our landscape the peculiar intimacy that distinguishes it from all others. As Richard Jeffries wrote in 1884, 'without hedges, England would not be England'.[76]

Between 1946 and 1974 more than one-quarter of all the hedgerows in England and Wales—approximately 120,000 miles—were destroyed by farmers in order to increase the efficiency of their land.

In response to increased public interest in countryside preservation, in 1981 Parliament approved the Wildlife and Countryside Act. It provided annual compensation to farmers if their land was to be managed less economically for environmental reasons. These payments were to be linked to "management agreements" and funded by local authorities, assisted by grants from the Countryside Commission. During a time of spending cuts, however, neither the local nor the central government has been able to commit sufficient resources to take advantage of this provision. Subsequently, a Labour peer, Lord Melchett, the president of the Ramblers' Association, called for extending the powers of local planning authorities to control significant changes in the countryside, as well as for legislation that would allow people greater access to the countryside. The Countryside Commission has also recommended that planning controls be established over farm buildings, farm roads, and afforestation in the uplands (land over 800 feet high) in England and Wales. The Thatcher government initially opposed substituting the current voluntary approach to conservation—which is based on compensating farmers whenever their freedom to do what they want with their land is restricted—with a formal system of planning controls. In 1984, however, the position of the government began to change, and it is likely that the 1981 legislation will be strengthened. Speaking at the Conservative party's annual conference, William Waldegrave, the junior environment minister, stated: "We are hammering out a new strategy for the countryside under which of course we must pay farmers properly for producing . . . food, but under which we can also recognize that it may cost money to save and renew the landscape, the wild animals, the birds and the flowers which we love."[77]

## THE EUROPEAN COMMUNITY

The British system of land-use planning, like the British approach to pollution control, has also found itself challenged by the European Community. While the first generation of Community initiatives in the area of environmental protection were primarily remedial in nature, the second action program, introduced in 1977, was preventive in focus; it sought to prevent the creation of pollution rather than simply to counteract its effects. In its third action program the Community explicitly focused on land-use planning. A draft report argued, "Land in the Community is a very limited and much sought after natural resource. The way in which it is used very largely conditions the quality of the environment. Physical planning is therefore one of the areas where a preventive environment policy is very necessary and beneficial." In 1980 the European Commission tabled a proposed directive "concerning the assessment of the environmental effects of certain private and public projects" which would have required each member state to "ensure that before any planning permission is given, projects likely to have a significant effect on the environment . . . be made subject to an appropriate assessment." The directive was then forwarded to the Council of Ministers. It has subsequently been the subject of much debate among officials from the member states as well as within a number of national legislatures.[78]

At hearings held by a committee of the House of Lords, the Department of Environment, the Association of Metropolitan Authorities, and the Confederation of British Industry all indicated their strong preference for an informal code of practice rather than a directive. While the Lords approved a report generally favorable to the directive, in a debate in the House of Commons the government indicated its formal opposition to its adoption—while supporting in principle the desirability of assessing the environmental effect of development projects.

The government is concerned that the adoption of a formal system of environmental impact assessments (EIAs) would deprive the British planning system of much of its flexibility. The current system enables each planning authority to determine for itself the extent to which it wishes to insist on strict standards of environmental quality for local developments. As a result, local planning authorities in some British communities have been both able and willing to accept somewhat higher pollution levels as the price of additional investment. One likely effect of the adoption of EIAs would be to make the environmental quality standards established by local authorities more uniform. This would be entirely in keeping with the underlying thrust of the commission's environmental policy, which is to encourage "harmonization" both among and within its member states. It would, however, represent a sharp departure from "the traditional pragmatic British approach of deciding each case on its merits."[79]

R. H. Williams writes: "There seems to be an underlying fear that the directive is the product of an alien system, and with it would come the imposition of fixed standards of environmental quality to be rigidly adhered to, and rigid procedures to be followed."[80] In both the Federal Republic of Germany and the Netherlands, for example, the determination on any application for permission to build is an administrative act: approval is automatic provided the proposal conforms with an approved development plan. Moreover, the directive would sharply conflict with the British government's recent commitment to streamline "a planning procedure, which it considers is already, for reason of over elaboration, causing delays." Thus, "just at the time when the British Government was trying to loosen its planning system—and succeeding to some extent—a certain tightening will...have taken place." Nigel Haigh, who has closely monitored the impact of the EEC on British environmental policy, concludes: "It will be one of the quirks of international policy making that a country with a mature planning system will have had it stiffened as the indirect result of legislation ill thought out by a country [France] which has no national planning system and has never believed in it anyway."[81]

The impact of the adoption of a formalized system of environmental assessment on British planning policies and practices remains unclear. In all likelihood it would improve the quality of information made available to local planning authorities and increase the amount of attention given to the environmental impact of major development projects undertaken by government departments, statutory undertakers, and nationalized firms. Both of these developments would be likely to strengthen the role of local and national amenity interests. There is no guarantee, however, that EIAs would ultimately affect the outcome of specific planning decisions, which would continue to be decided by either local planning authorities or the central government in terms of whatever criteria they chose to employ.

In fact, environmental impact assessments are steadily becoming more common in Britain. Over the last decade both local planning authorities and central government departments have commissioned detailed studies of the impact of particular developments—most frequently in the area of energy—on the surrounding community. The Department of the Environment has also prepared a detailed manual for the environmental assessment of proposed large-scale industrial developments and has urged developers, local planning authorities, and central government agencies to make use of it. They are not legally required to use it, however, and many have not done so: planning authorities continue to enjoy substantial discretion in determining whether or not they wish to require an environmental impact statement, and if so, what it should include. But to the extent that this trend continues, Britain will have moved toward the adoption of a system of environmental impact assessments, though without making them either mandatory or uniform.

CONCLUSION

In addition to having the world's oldest pollution-control agency, Britain has the world's most comprehensive system of land-use planning. Because the decisions made under the latter regulatory system provide extensive opportunities for public participation, Britain's extensive array of amenity and community organizations has been able to exercise an important influence on Britain's land-use policies. The result is apparent to anyone who has visited the British countryside.

But while the politics and administration of land-use planning in Britain more closely resemble those of environmental policy making in the United States than does the British system of pollution control, the two systems have more in common than either does with the pattern of environmental regulation in the United States. Both sets of policies are relatively decentralized, encourage negotiation and compromise, make little use of fixed standards, allow substantial administrative discretion, minimize the use of both lawyers and courts, and in the final analysis appear to strike roughly the same balance between amenity and economic values and between national and local interests. The significance of these contrasts is the subject of Chapter 4.

# CHAPTER FOUR

# A Comparison of British and American Environmental Regulation

While it is difficult to determine the comparative effectiveness of governmental regulations in different countries, Great Britain and the United States appear to have made comparable progress in improving air and water quality, preserving scenic areas, and protecting the health and safety of their people. On balance, neither nation's regulatory policies have been significantly more or less effective than the other's: both have had some notable achievements and some conspicuous failures. The British system's emphasis on secrecy, informality, and voluntary compliance has proved no less effective in controlling the externalities associated with industrial growth than the more open, legalistic, and adversarial style of regulation adopted in the United States. Moreover, both nations appear to have struck a roughly similar balance between economic and amenity values: each has allocated approximately the same proportion of its gross national product (GNP) to pollution control. And in neither nation have environmental controls measurably contributed to the difficulties its economy has experienced in recent years.

In most areas of environmental regulation, not only are American laws and regulations stricter, but nonindustry constituencies enjoy substantially more access to the regulatory process. Why, then, is the American record not demonstrably better? And why have the burdens on industry not been greater? The main reason is that the enforcement of American laws and regulations has been uneven. When one focuses on the *implementation* of environmental regulation rather than on the regulations themselves, the significance of many of the differences in the two nations' environmental policies diminishes.

But while the two nations' distinctive strategies for regulating industry have differed only marginally in their environmental and economic impact, they produced markedly divergent political outcomes. Environmental policy in the United States has been associated with a major increase in political conflict between industry and government; it has made an already adversarial relationship significantly more contentious. This has not been the case in Britain. While the relations between business and government in this policy arena have certainly not been free from conflict, on the whole the British business community has been relatively satisfied with its nation's system of environmental controls. These different responses are in large measure due to the different ways in which environmental regulations have been made and enforced in the two societies.

## THE EFFECTIVENESS OF REGULATION

### Methodology

Measuring the impact of regulation on environmental quality is extremely difficult within a given political system, let alone cross-nationally. One problem is that governmental regulation is only one of several factors that affect both emission levels and environmental quality. Without doubt the most important factor is simply the rate of economic growth: all other things being equal, the more industrial production, the more pollution. (Between 1965 and 1975 total industrial production increased 31 percent in the United States and 14 percent in Britain.)[1] On balance, the global slowdown in growth rates since 1973 has done more to improve air quality in both Great Britain and the United States than the regulatory efforts of either the Alkali Inspectorate or the EPA. For many communities in the heavily industrialized regions of both countries, undoubtedly the most important single factor responsible for reducing the pollution to which they have been exposed over the last decade has been the rate at which plants have been closed in response to international competition and a decline in domestic demand. (Between 1965 and 1975 iron and steel production declined by 28 percent in Great Britain and 10 percent in the United States.)[2] Slower growth rates also have a positive effect on the preservation of wilderness and other pristine areas: the less growth, the less industry's need for new sites for industrial production, energy generation, and the mining of natural resources.

The emissions of particular pollutants are affected not only by the volume of industrial production but also by its particular mix. (As of 1970, high polluting industries constituted 14.9 percent of the GNP of the United States and 12.4 percent of Britain's GNP.)[3] Since new plants are in general less polluting than older plants, to the extent that older facilities either modernize, relocate, or shut down, emissions are likely to decline. However, the role of environmental regulation in changing the industrial base of either country appears to be marginal; economic factors are far more important.

While pollution is controlled primarily by regulation of emissions, emission levels represent a proxy for what is the primary purpose of pollution control, namely, the improvement of air and water quality. The relationship between the two, however, is by no means straightforward. Water quality is significantly affected by the presence of various chemicals in the runoffs from agricultural lands—a form of pollution beyond the control of water-pollution authorities. Air-quality levels are influenced not simply by the volume of emissions but also by the strategy by which they are controlled. The decision of pollution-control authorities in Great Britain and the United States to allow utilities and factories to disperse emissions has led to the improvement of air quality in many regions, although emission levels have not been proportionately reduced. For other areas, of course, the reverse is the case: their emission levels can be reduced and yet their air quality can continue to deteriorate. This is a particularly important issue for both the United States and Britain, whose utilities produce considerable quantities of sulfur dioxide that are then displaced both within their boundaries and outside them.

Topological factors also affect environmental quality. Because Great Britain has a relatively large number of rapidly flowing rivers, a proportionately long coastline, and relatively high and strong winds, much of its environment is capable of absorbing relatively large amounts of pollution without adverse affects on the environmental quality experienced by its citizens. Certainly no British city has the peculiar geographical and climatic disadvantages of Los Angeles. On the other hand, Britain's much smaller size and greater population density (288 persons per square kilometer versus 22 in the United States) provide it with fewer options for dispersing its heavily polluting industries away from population centers.[4] They also make the preservation of land for conservation purposes more difficult.

In both countries different regions experience different levels of environmental quality, owing both to topographical factors and to the pattern of industrial location. Moreover, the policies of each government have emphasized the improvement of environmental quality in some regions more than in others. As a result, environmental quality varies substantially within each nation; in fact, the levels of both air and water quality vary as much within Great Britain and the United States as between them.

Comparing the two countries in terms of the effectiveness of environmental regulation is also complicated by a lack of consensus as to what constitutes a reasonable risk or unacceptable hazard. Consider, for example, the following comparison of American (in this particular case, California) and British regulations governing the siting of liqufied natural gas terminals:

> Recently California and the United Kingdom have approved sites for Liquefied Energy Gas (LEG) terminals. In this, and perhaps this alone, they are the same. After a long drawn-out process in which it proved impossible to approve any of the proposed sites, California finally, with the help of a new statute passed expressly for the purpose, was able to give approval for an LEG facility at the remotest of all the sites on the list of possibles: Point Conception. Scotland has a longer coastline than California and most of the country is very sparsely populated (less than 25 persons to the square mile) and yet the approved site, at Mossmorran and Braefoot Bay on the Firth of Forth, lies within the most densely populated part of the entire country (with a population density of between 250 and 500 persons per square mile). Moreover, laden tankers will pass within a mile or so of Burntisland (an industrial town) and sometimes within four miles of Edinburgh—the capital city of Scotland! If the California siting criteria (explicit in Statute 1081) were to be applied to the Scottish case it would be quite impossible to approve the Mossmorran/Braefoot Bay site, and if the United Kingdom criteria (implicit in the Mossmorran/Braefoot Bay approval) were to be applied to the Californian case, any of the suggested sites could be approved, which means that the terminal would go to the first site to be suggested—Los Angeles harbor.[5]

As we shall subsequently see in more detail, this illustration is not atypical: American regulations in the area of health and safety have frequently been significantly stricter than Britain's. But this does not mean that American environmental regulations are necessarily more effective: whether they are or not depends on one's assessment of both the seriousness and the probability of the harm that each nation is attempting to prevent. If one accepts the risk assessment of the Health and Safety Inspectorate, then the stricter American standard did not provide any additional protection to the residents of California; it merely added to the difficulty and expense of securing siting approval. On the other hand, if one accepts the risk assessment used by the State of California—which was actually based on the results of the least conservative of six separate studies, each of which reached a different conclusion—then the American standard is more effective; that is, it provides better protection.[6]

In this context, it is worth noting that while the Americans have permitted nuclear power plants to be constructed near major population centers, almost all of Britain's nuclear power facilities are located in relatively sparsely populated areas. Evidently the two nations' regulatory authorities have appraised the relative risks of liquefied natural gas (LNG) and nuclear power plants differently. More generally, to the extent that environmental regulation seeks to avoid future harms rather than to ameliorate current ones, comparative assessment of the effectiveness of regulation is extremely difficult. In many cases there is simply no objective of the nature or magnitude of the hazards one is seeking to minimize.

An important purpose of environmental regulation is to protect the health of the public. From this perspective, assessments of the severity of restrictions on particular substances, or even their concentrations in either the surrounding environment or the human body, merely represent proxies for what ultimately counts—how long people live and how healthy they are while they are alive. Such statistics are readily available. The life expectancy of an American born in 1981 is estimated to be 75 years; for an individual born in Great Britain, 74 years.[7] The infant mortality rates of both societies are now identical. Such statistics, however, tell us relatively little about the effectiveness of each nation's system of environmental regulation. For obviously many other factors affect both longevity and disease rates, including the availability and quality of health care, the quality of public sanitation, personal habits and lifestyles, and, most important, GNP.

Age-adjusted cancer rates are higher in Britain than in the United States: in 1979, 163.7 deaths per 100,000 males were due to malignant neoplasms in the United States compared to 185.4 for Britain; the comparable statistics for females are 107.5 for the United States and 122 for Great Britain. During the 1970s American age-adjusted cancer rates for men increased somewhat, while in Britain they slightly declined. Female age-adjusted cancer rates increased in both countries, though somewhat more in Great Britain.[8] These statistics, however, tell us only about the recent past. Since many potentially dangerous substances have been introduced into the environment in the last few decades, it may be some time before their full effect is felt. Thus it may not be possible to evaluate the relative effectiveness of many of the distinctive regulatory policies adopted by the United States and Great Britain until the end of the century.

The precise relationship between industrial activity and cancer rates remains a source of considerable controversy, however; it is by no means clear what proportions of cancer are due to industrial production in general and to pollution in particular. (In both countries the most important cause of cancer remains cigarette smoking, which neither government has made a serious effort to restrict.) The fact that cancer rates differ so substantially between the sexes in both countries—British women have much lower cancer rates than American men—suggests that environmental policy itself has had only a small effect on the incidence of cancer, since presumably environmental controls—or the lack thereof—affect men and women equally. The relative importance of factors other than industrial activity is also suggested by the fact that the residents of England and Wales have a lower age-adjusted cancer rate than do American blacks.[9] When we turn from cancer to other diseases caused by industrial activity, establishing the causality necessary to measure the effectiveness of government regulation is equally complex and controversial.

There is yet another difficulty with cross-national evaluations of the effectiveness of environmental regulation: each nation has its own priorities. The issue of environmental regulation has been cast in terms of the protection of public health to a far greater extent in the United States than in Britain. Not only is the United States the only nation to establish an entirely separate system for the regulation of carcinogens, but in general the American public appears to be far more anxious than the British—or indeed the public of any other industrial nation—about the threats to health and safety posed by industrial products, production, and waste disposal. In Great Britain, largely because of its greater population density, conservation and land-use issues appear much more salient. British planning policy has also emphasized visual amenity to a far greater extent than have zoning regulations in the United States.

Since 1970 air pollution has been the most salient environmental problem only in the United States and Japan, though it has recently assumed greater importance in Germany; in Great Britain and in the rest of Europe, improving water quality has been accorded a higher priority over the last two decades. The appropriate weight of trucks came to be defined as an environmental issue in Great Britain but not in the United States. In America the pollution generated by mobile sources has occupied a central place on the environmental agenda, while in Great Britain, with the exception of lead, there has been much less public interest in curbing motor vehicle emissions. The American environmental movement has given high priority to the protection of endangered species of animals. In Britain this issue has somewhat less importance for the simple reason that there are fewer wild animals left to protect, although environmentalists in both societies have devoted considerable efforts to halting the hunting of whales and protecting birds. British naturalists appear to be especially interested in the protection of various species of grasses and flowers while American environmentalists have accorded a similar priority to the preservation of wetlands. In Great Britain, as in the other member states of the European Community, the regulation of noise has emerged as an important component of environmental policy; in America this issue has little political salience and is handled primarily at the local level. The protection of historic buildings and landmarks has traditionally been a concern of British environmentalists, while the American environmental movement has tended to leave this issue to other interest groups.

Such examples can be multiplied indefinitely. Each enormously complicates the task of evaluating policy effectiveness. At least a part of the differences in the environmental regulations of the two countries stems from the fact that they have both different problems and different priorities.

There is another obstacle to evaluating the effectiveness of environmental regulation in the two societies: the lack of a comparable time frame. While environmental controls in both nations have existed throughout the twentieth century, the most important contemporary initiatives in British environmental regulation took place during the 1940s and 1950s: the Town and Country Planning and Countryside acts, both enacted shortly after World War II, and the passage of the Clean Air Act and the expansion of the jurisdiction of the Alkali Inspectorate, both of which occurred during the 1950s. Only with respect to the control of water pollution did any major policy departure occur during the 1970s, and that was primarily administrative in nature.

In the United States, by contrast, virtually all of the important policy initiatives during the postwar period have taken place since the late 1960s. Between 1969 and 1972 the United States enacted the National Environmental Policy Act, the Clean Air Act Amendments, and the Federal Water Pollution Control Act, all of which radically transformed the scope and enforcement of American environmental controls. Moreover, throughout the 1970s in America, in contrast to Britain, environmental regulation has remained in a considerable state of flux: eight major laws were either enacted or substantially amended between 1970 and 1980.

As a result, comparative evaluation of policy effectiveness throughout the postwar period is "unfair" to the United States, since many of its most important environmental regulations only began to be implemented in the mid-1970s; to use 1970 as a base point, on the other hand, is to minimize the progress made in Britain, since many of its regulations began to take effect before that date. A further complication is that while

British data go back to the 1950s, data on pollution in the United States are extremely fragmentary before 1970. In addition, Britain appears to have a more extensive and accurate system for monitoring pollution levels than does the United States. Fifteen years after the establishment of the EPA, the United States still has no comprehensive system for actually measuring changes in either air or water quality.

In view of these difficulties, any conclusions about the relative effectiveness of British and American environmental regulations must be drawn with caution. The evidence we have available does suggest, however, that both nations have made measurable though uneven progress in reducing pollution levels, safeguarding public health, and preserving amenity values. Of equal importance is what the available data do *not* demonstrate, namely, that either nation's environmental policies have been significantly more or less effective than the other's.

*Air Pollution*

Since the beginning of the Industrial Revolution, Great Britain's most serious air-pollution problem has been the smoke produced by the burning of coal.[10] Not only did emissions of particulates impair visibility, but smoke concentration levels presented a serious threat to public health. Between 1958 and 1981, however, smoke emissions from domestic coal combustion declined by more than 80 percent; industrial emissions declined even more dramatically, from 0.51 million tons in 1958 to 0.03 in 1981. Average urban ground-level concentrations of smoke declined from 150 milligrams per cubic meter in 1957 to about 30 milligrams in 1978; by 1980–81 they were one-eighth of their 1960–61 levels. Periodic high pollution episodes still do occur as a result of adverse climatic conditions, but they have steadily diminished in both frequency and intensity; no excess deaths have been attributed to concentrations of smoke in Great Britain since the early 1960s. As of 1970 "there was no longer any evidence of sharp increases in illness related to pollution."[11]

The increased use of smokeless fuels cannot be attributed entirely to government regulation. Changes in social mores, leading British housewives to view a coal fire in the living room as a source of dirt rather than comfort, the installation of central heating in newly constructed council houses, and the increased availability of natural gas also played an important role in the switch to smokeless fuels. Nonetheless, the 1956 Clean Air Act does seem to have played an important role. Those communities in South Wales, the East Midlands, and sections of northern England that have chosen not to establish smoke-control zones continue to experience relatively high smoke concentration levels; the improvement in air quality has been dramatic in London, which has done so. Between 1956 and 1966 smoke emissions declined by 76 percent in London, but only 20 and 30 percent in the Midlands and the the north of England.[12] It turns out that the famous London fog, immortalized by Sir Arthur Conan Doyle and other British novelists, was in fact produced by the burning of domestic coal (also immortalized by Conan Doyle and numerous other British novelists). Since the early 1960s it has all but disappeared. In fact, "the Clean Air Act has arguably been responsible for the biggest single improvement in the lives of Londoners in the last twenty-five years."[13] One scientist notes:

> It is encouraging...to see that the urban excess of bronchitis mortality shows signs of reduction, and in particular death rates from bronchitis are no higher in London, at least up to about age 65, than in rural areas. It may be a little early to promote London as a health resort, but certainly there have been major strides in the control of pollution that have made it both a healthier and pleasanter place to live in than it was earlier this century.[14]

Since the mid-1960s the number of bird species in London has doubled and the number of delicate plants able to survive in the city has substantially increased. Eric Ashby and Mary Anderson write: "Before the 1950's London was a dirty city with high levels of smoke and sulfur dioxide. Today the air over London is as clean as the air over East Anglia."[15] To take a more international comparison, in 1967–68 the concentration levels of suspended particulate matter (smoke) averaged 55 micrograms per cubic meter in London, 112.5 in New York City. The major industrial city of Sheffield used to be regarded as "smoky and dirty"; thanks to its dramatic reductions in the levels of smoke emissions, Sheffield now claims to be "the cleanest industrial city in Europe."[16]

In contrast to smoke emissions, most of which are produced by individual households, sulfur dioxide emissions are generated primarily by power stations and other stationary fuel-burning sources. Sulfur dioxide emissions in Great Britain remained fairly stable through the 1960s and early 1970s, averaging approximately 6 million tons. Between 1970 and 1980, however, aggregate sulfur dioxide emissions decreased by 24 percent. (While the overall trend is downward, weather conditions produce some yearly variation.) Emissions of sulfur dioxide have also declined in relation to the units of energy consumed: from slightly more than 30 tons per 1,000 tons of energy consumed in 1965 to approximately 26 tons in 1975.

Urban ground concentration levels of sulfur dioxide have declined more dramatically. This improvement appears to be attributable to two factors. One is the government's "high chimney policy" of ejecting waste gases from utilities and other large plants at high velocity into airstreams up to a kilometer high away from Britain's population centers, thus dispersing them out to sea. The second has been the reduction in smoke emissions. It turns out that when the smoke blanket is eliminated, less sulfur dioxide remains trapped at ground level. (This has been an un-anticipated benefit of the 1956 Clean Air Act.) Thus urban ground concentration levels of sulfur dioxide fell from 150 micrograms per cubic meter in 1958 to about 70 micrograms in 1978, while daily concentration levels at urban sites during the winter months declined from 70 in 1977–78 to 49 in 1980–81. Both London and the northwest of England (which includes Manchester and Merseyside) recorded a 30 percent reduction in the annual mean daily concentrations of sulfur dioxide between 1970 and 1976.

Emissions from numerous major stationary sources of pollution have also declined, largely as a result of the transfer of responsibility for regulating them from local authorities to the Alkali Inspectorate. Cement works, for example, emitted 16 tons of particulate per 1,000 tons of production in 1958; by 1974 their average emissions had declined to 1.5 tons. Coal-fired power stations emitted 23 tons of particulate emissions for every 1,000 tons of coal burned; by 1974 this figure had declined to 3 tons. In addition, "massive improvements have been made in the steel industry; the mustard cloud that used to pall steel towns like Port Talbot, home base of British Steel Company's Wales division, has gone."[17] The Alkali Inspectorate has encouraged firms in a number of industries to switch to less polluting modes of production. In 1958, for example, 295 pottery works in Stoke-on-Trent were using smoke-producing methods of firing; a decade later all had installed oil, electric, or gas-fired ovens. Between 1958 and 1968 the number of gas and coke ovens fell from 477 to 277, liquid-gas works from 140 to 81, and benzine works from 247 to 124. During the 1970s lead emissions from plants registered with the Alkali Inspectorate were halved, arsenious oxide emissions were reduced by nearly two-thirds, and emissions of vinyl chloride monomer from PCV plants declined by more than three-quarters.

Since 1972 Great Britain has progressively restricted the amount of lead permitted in petrol. As a result, even though gasoline consumption increased from 15.9 million tons in 1972 to 18.72 million tons in 1981, vehicular emissions of lead declined from 8.1 thousand tons to 6.7 thousand tons. On the other hand, both carbon monoxide and nitrogen oxide emissions from motor vehicles, which have not been regulated, have increased since 1970, primarily as a result of greater automobile use.

Like Britain, the United States has made measurable progress in reducing emissions and ground concentration levels of both particulates and sulfur oxides.[18] Ground concentration levels of sulfur dioxides fell 60 percent between 1966 and 1970 and an additional 40 percent between 1970 and 1979. Although the burning of coal by electric utilities—the main source of sulfur dioxide—increased by 85 percent between 1971 and 1981, the actual emissions of sulfur dioxide registered a slight decline as a result of both the installation of scrubbers and the increased use of low-sulfur coal. Between 1970 and 1977 particulate emissions declined even more substantially, from 25.5 to 13.8 million tons, while ambient concentration levels of this pollutant were reduced from 72 micrograms per cubic meter to 60. Consequently by 1980 almost all air-pollution control regions had met the EPA's primary standards for these two hazardous pollutants.

While Britain's efforts to reduce air pollution have focused primarily on the domestic burning of coal, the United States has given highest priority to reducing emissions from automobiles. In the decade following the Clean Air Act Amendments of 1970, motor vehicle emissions of carbon monoxide declined from 76.7 to 49.4 grams per mile traveled while total carbon monoxide emissions declined by 23 percent. The average concentration of carbon monoxide in urban areas declined by one-third between 1972 and 1978 while overall concentration levels fell an average of 31 percent between 1975 and 1980. For hydrocarbons, which play an important role in the formation of photochemical oxides (ozone), the reduction has been more substantial, from 10.8 to 5.3 grams per mile between 1970 and 1980. Although the reduction in the average size of vehicles following the energy crisis also played a role, these improvements result primarily from the installation of catalytic converters in automobiles. As in Britain, however, nitrogen oxide emissions from mobile sources have actually increased. Finally, as a result of the increased use of unleaded gasoline, air-quality data from 92 urban monitoring sites in eleven states revealed a 64 percent decline in average ambient concentrations of lead between 1977 and 1980.

The controls over both stationary and mobile sources have produced measurable improvements in air quality for most urban residents.[19] In twenty-three metropolitan areas the number of "hazardous days" declined from 1.8 to 0.13 between 1974 and 1981; the number of "very unhealthful" days was reduced by 65 percent and the number of "unhealthful" days by 50 percent. As in Britain, however, improvements in air quality have not been uniform. The nation's two largest urban areas continued to experience substantial pollution problems: between 1976 and 1978 New York and Los Angeles each averaged more than 200 unhealthful days per year. The air quality of Houston, Chicago, and Kansas City deteriorated during the second half of the 1970s. Overall, pollution levels declined in nine metropolitan areas between 1974 and 1978, increased in eight, and were essentially unchanged in six.

While the regulation of automobile emissions by the federal government since 1967 has contributed to improved air quality in many metropolitan areas, the actual impact of the controls established by the Clear Air Act Amendments of 1970 over stationary sources is somewhat less clear. Admittedly fragmentary data collected by the EPA indicate that sulfur dioxide concentrations actually declined more each year between 1964 and 1971 than they have done since 1971. During the second part of the 1960s they fell 11.3 percent per year, while during the 1970s they declined by only 4.6 percent. Similarly, particulate concentration levels declined 2.3 percent per year between 1960 and 1971 but only by 0.6 percent per year between 1972 and 1981. Robert Crandall argues that "these data suggest that pollution reduction was more effective in the 1960's, before there was a serious federal policy dealing with stationary sources, than since the 1970 Clean Air Act Amendments."[20] Moreover, as in Britain, some share of the decline in both sulfur dioxide and participate emissions and concentration levels during the 1970s must be attributed to the decline in energy consumption.

## Water Pollution

Both nations have experienced considerably more difficulty in improving water quality, largely because of the substantial expenses involved in installing water-treatment facilities. On balance the British effort appears to have been somewhat more successful.[21] In 1958, 86.1 percent of the lengths of rivers in England and Wales were sufficiently clean to support varied fish life and to be suitable for drinking after treatment; by 1975 this figure had increased to 91.4 percent. Approximately 12.5 percent of the length of all tidal and 6.2 percent of the length of all nontidal rivers had their quality measurably improved during this period. Between 1958 and 1980 the length of "grossly polluted" and "poor quality" nontidal rivers and canals declined by nearly half, from 4,520 kilometers to 2,810, while that of polluted tidal rivers decreased from 760 kilometers to 440 kilometers. North Sea energy development has played an important role in this improvement. With the increased availability of gas and oil from the North Sea, Britain has been able to close its coal gasification plants, thus reducing the emissions of phenols and other chemicals into waterways.

During the 1950s Britain's companies were using more than 40,000 tons of a branched alkyl benzene sulfanite to make domestic detergents each year.[22] The result was the poisoning of fish and water plants, as well as the creation of massive quantities of foam that clogged up sewerage plants; one was actually buried under fifteen feet of foam for several weeks. In 1957 a Standing Technical Committee on Synthetic Detergents was established, and in 1964 it secured a voluntary agreement by manufacturers not to market "hard" detergents for domestic use. Within seven years 95 percent of the detergents available for sale in Britain were biodegradable, and water quality markedly improved.

As in the case of smoke concentrations, national statistics on water quality conceal considerable regional variations. Some estuaries in England and Wales, including the Tyne, Wear, Humber, Mersey, and Severn, remain heavily polluted. "In many cases the level of pollution is sufficient to end the commercial use of shellfish, affect birds and sea mammals and restrict recreational use of nearby beaches."[23] The contamination of urban water supplies by agricultural pesticides remains a serious problem in many areas. The seepage of hundreds of gallons of pesticides into the Rhyl River in Yorkshire in 1978 threatened to contaminate York's water supply, and nitrate concentration levels have occasionally made it necessary to recommend the use of bottled water for infants in East Anglia. In the three British rivers surveyed by the OECD, the Lee, the Wear, and the Irwell-Mersey, the annual mean concentrations of nitrates increased by more than 50 percent between 1965 and 1975. In addition, the discharge of untreated sewage into the seas around Britain remains a serious problem: of the nation's 633 bathing beaches, 190 exceed the European Community's standards for bathing water quality.

The most noticeably improved river in Great Britain is the Thames. By 1950, after nearly a century of neglect, the Thames was, for all practical purposes, dead, unable to support any marine life. In the June 1978 issue of *The Environment*, Trevor Holloway reported that "the lower Thames was so heavily polluted for several months of the year that no dissolved oxygen could be detected . . . and a disquieting smell rose from the river. . . . Today, less than three decades later, the Thames is rated as the cleanest tidal river in the world."[24] The river's pollution level has dropped nearly 90 percent since 1950, its offensive odor has disappeared, it currently contains more than 100 species of fish, and it is suitable for swimming. In September 1983 a salmon was caught in the river outside London, the first in more than 150 years. The fisherman was awarded a prize of £375, offered a decade earlier by the Thames Water Authority.[25]

According to data collected by the U.S. Geological Survey, the quality of surface waters in the United States changed relatively little between 1974 and 1981. The Survey's National Ambient Stream Quality Accounting Network records changes in concentration levels of five pollutants: dissolved oxygen, bacteria, suspended solids, dissolved solids, and phosphorus. "For each of these, the vast majority of monitoring stations show no significant change in pollutant concentrations . . . with those showing trends of increases balanced by a comparable number of decreases."[26] (This does not mean, however, that regulation has had no effect, since given the significant increase in industrial output in the United States during this period, without controls water quality would undoubtedly have deteriorated.)

These statistics, like Britain's, conceal important regional variations. Numerous rivers, such as the Hackensack in New Jersey and the Pemigewasset in New Hampshire, have literally been brought back to life, and substantial progress has been made in improving the quality of the Great Lakes, especially Ontario and Erie; the latter, which was once considered dead, is now teeming with fish. Moreover, "Atlantic salmon have returned to New England's Connecticut and Penobscot rivers and shellfish-bed public-beach closings due to dangerous bacteria levels have become less frequent."[27] Overall, the quality of approximately fifty major bodies of water improved considerably during the 1970s. But according to the EPA, while "technology-based water pollution controls have improved the quality of many of the nation's rivers and streams . . . point and nonpoint sources of pollution, as well as other factors, continue to cause violations of water quality standards and are limiting water uses in many areas of the country." Two-thirds of the states reporting trend information to the EPA in 1982 indicated "generally improving water quality trends," but another third reported no progress at all.[28] The water quality of the streams and estuaries of the United States as a whole has improved slightly while that of its lakes and reservoirs has marginally deteriorated.

### Other Comparisons

Of the major oil spills—defined as those in excess of 2,000 tons—in marine waters recorded by the OECD between 1967 and 1978, eight occurred off the coast of the United States, while four, including the largest—the grounding of the *Torrey Canyon*—occurred in British territorial waters. Between 1973 and 1977 the production of PCBs declined in Great Britain from 4,067 to 283 tons; in America the percentage decline was significantly less—from 19,132 tons in 1973 to 6,046 in 1977—though the absolute decline was much greater. Between 1967 and 1975 the consumption of commercial fertilizer per ton of cropland doubled in the United States while it increased only 25 percent in Great Britain. On the other hand, the absolute level of consumption is significantly higher in Britain (26.4 metric tons per kilometer of cropland) than in the United States (9.9 metric tons). Both countries registered a comparable increase in the use of fertilizers containing nitrogen from 1965 and 1975, though the intensity of British use was substantially greater— in fact, more than triple that of the United States. Between 1970 and 1979 the concentration of DDT in human tissue in the United States declined from 8.07 parts per million to 3.64 and that of Dieldrin from 0.23 to 0.11.[29] While we lack comparable British data, the average daily intake of both DDT and dieldrin in Britain has remained substantially below that judged acceptable by the World Health Organization.[30]

The United States' safety record with respect to nuclear power appears substantially better than Britain's. "One of the worst atomic accidents in history" occurred at Britain's nuclear fuel reprocessing plant at Windscale in 1957, producing 40 to 400 times as much radiation as was released at Three Mile Island in 1972; the government has estimated that this radiation may have been responsible for as many as 250 cases of thyroid cancer.[31] (In the case of these and other catastrophes and near-catastrophes that periodically beset both countries, though, it is difficult to know how much to attribute to inadequate regulation and how much simply to bad luck.) In November 1983 a British television program titled "Windscale in the Nuclear Laundry" reported that the dumping of radioactive wastes into the Irish Sea was causing cancer rates in the surrounding community well above the national average. Although a subsequent government investigation did not find sufficient evidence to support this contention, controversy over the safety of the reprocessing facility continues.[32]

Both nations appear to be experiencing considerable difficulty in enforcing their regulations governing the disposal of toxic wastes. Britain enacted legislation regulating toxic waste disposal in 1972, the United States in 1976. Each appears to be confronted with a considerable number of sites at which toxic wastes have been improperly disposed of—in many cases threatening to contaminate groundwater supplies—though to date America's problems appear more serious. A 1975 survey in Great Britain revealed a total of 51 disposal sites that could be considered a risk to underground water supplies, while estimates of the number of waste disposal sites that may pose "a significant threat to public health and/or the environment in the United States" range from 16,000 to 27,000; the EPA has identified 115 sites as being of highest priority for remedial action under its Superfund program because of the risk of air, surface water, or groundwater contamination, the toxicity of the wastes, and the potential for human exposure.[33]

On the other hand, each nation appears to have been able to bring on stream major new supplies of energy—Great Britain in the North Sea off Scotland and America in the North Slope of Alaska—without producing any serious environmental damage, although because of the extreme fragility of the Alaskan tundra, the American achievement was more difficult. (In fact, the same company was extensively involved in both projects: British Petroleum.) And both nations have been equally hesitant to address the problem of acid rain: the criticisms from Canada and the American northeast of the lack of effective controls over sulfur dioxide emissions from coal-burning utilities in the Midwest precisely parallel those made of English emissions controls by Sweden and the Scots.[34]

There is no question, however, that the British system of land-use planning has been, on the whole, extremely successful in reconciling amenity and economic values; it can claim much of the credit for the relative lack of urban sprawl in Great Britain, as well as for the generally pleasing appearance of British towns and much of the British countryside.

> The vaunted English development control system ... has surely been suc-
> cessful in doing what it was asked—controlling development in a country
> with a population of about 55 million in an area about the size of Wisconsin.
> One can only marvel at the absence of sprawl and the resulting open
> countryside and coastline.[35]

Since zoning is primarily a local responsibility in the United States, the American record in this regard is much more uneven and, on the whole, much poorer—though a few states have imposed highly effective land-use controls.

The amount of land designated as national wilderness, excluding Alaska, more than doubled between 1970 and 1980, while 75 units, totaling more than 2.5 million acres, were added to the National Park Service. Likewise, the size of the greenbelt surrounding British towns and cities doubled during the 1970s and additional acres have been acquired by the National Trust. According to the International Union for the Conservation of Nature and Natural Resources, the United States and Great Britain are among only a handful of countries more than 5 percent of whose land is classified as conservation areas.

My argument is *not* that either the British or the American government has been effective in safeguarding the health of the public and protecting the quality of the ecosphere. Whether the effort devoted to environmental protection in either nation has been adequate in the face of the manmade hazards confronted by its inhabitants is an issue on which reasonable people disagree; it is also beyond the scope of this study. Rather, my contention is that there is no evidence that either nation's policies have been particularly more or less effective: that is to say, depending on one's point of view, they have been equally effective or equally inadequate. Somewhat ironically, this conclusion is echoed by many of the more vocal critics of British environmental policy. In questioning the general sense of satisfaction of British government officials with their nation's approach to environmental regulation, they have not contended that environmental regulation is any less effective in Great Britain than in the United States, only that it is equally ineffective.[36]

## IMPLICATIONS

At first glance, this conclusion about the relative effectiveness of British and American environmental regulation is surprising. For, if we compare the actual laws and regulations of the two countries, the controls on industry enacted in the United States are certainly far stricter. While British environmental policy has remained incremental in orientation, the environment laws adopted by the United States in the late 1960s and early 1970s established extremely ambitious goals. The 1972 amendments to the Clean Water Act required all of the nation's waters to become "fishable and swimmable" by July 1, 1983; they also established a national goal of zero discharge of pollution into the nation's waterways by 1985. In addition, the Environmental Protection Agency was required to develop "pollutant-specific effluent standards to be applied to all industrial categories regardless of technological or economic achievability."[37] This legislation also established a set of strict timetables for the achievement of national emission standards: industrial discharges were required to employ "the best practicable technology" by July 1, 1977, and the "best available technology" by July 1, 1983.[38]

The Clean Air Act Amendments of 1970 were equally rigorous. The EPA was given 120 days to promulgate national air-quality standards "based on such criteria and allowing an adequate margin of safety [as were] requisite to protect the public health." In developing these "primary standards" the EPA was explicitly enjoined from taking into consideration the costs of meeting them. The states were then required to submit implementation plans that would enable these standards to be met by 1975. Section 112 required the EPA to regulate particularly hazardous pollutants to zero risk with an "ample margin of safety" regardless of cost.[39] This legislation also established the first "technology-forcing" pollution-control standard: motor vehicles were required to reduce their emissions of hydrocarbons and carbon monoxide by 90 percent within five years and of nitrogen oxides by 90 percent within six years, even though these "targets were acknowledged by all to be beyond the existing technological capabilities of the automobile manufacturers."[40]

Severe penalties were established for noncompliance. "Violation of an implementation plan, a new source performance standard, or a hazardous emission standard [was] punishable by a fine of up to $25,000 a day and one year in prison," while a "violation of EPA's motor vehicle fuel standards [was] punishable by a fine of up to $10,000 a day." Violators of the government's water-pollution standards likewise could be fined "up to $25,000 a day and sentenced to one year in prison."[41] The EPA was also given the power to issue abatement orders if public health was threatened. And in an attempt to prevent industry from dominating the enforcement process, the Clean Air Act gave private citizens the right to sue the EPA and provided for public hearings before standards could be issued.

British pollution-control statutes, by contrast, contain no deadlines or technology-forcing standards and omit any explicit reference to public health. The British "best practicable means" standard implicitly takes into account the costs of compliance, while its closest American equivalent, "best available technology," requires the installation of technology most likely to bring about maximum abatement regardless of costs or environmental necessity. Unlike the United States, Britain makes extremely limited use of either emissions or environmental quality standards, and its penalties for noncompliance remain modest. It does not require the holding of hearings before pollution-control standards are issued, nor are its citizens allowed to sue pollution-control authorities. It has also imposed fewer controls on automobile emissions. While the National Environmental Policy Act of 1969 requires detailed environmental impact assessments for all projects supported by the federal government likely to have a significant effect on the environment, the preparation of environmental impact statements has remained optional in Great Britain. Both the Countryside Act (1968) and the Water Act (1973) do require public bodies to take amenity considerations into account when formulating development proposals, but they do not mandate any specific procedures for doing so and compliance is not subject to judicial review. Although development controls are more extensive in Great Britain than in the United States, British policies in regard to land use are in general more flexible: Britain permits the commercial development of land in its national parks, while mineral exploration is strictly prohibited in national parks as well as in other conservation areas in the United States.

The Environmental Pesticide Control Act of 1972 required the registration of all 35,000 pesticides then on the market and established strict criteria regarding their potential health hazards. Pesticide use in Britain, by contrast, is subject to a voluntary screening program administered by the chemical industry itself; there are no statutory standards for risk assessment. The Surface Mining Control of Reclamation Act of 1977 established sixteen specific performance standards regarding the restoration of land used for strip mining. In Britain the National Coal Board negotiates separately with each property owner regarding compensation for the use of his or her land, and voluntary agreements have been reached in all but a handful of cases.[42]

## THE ENFORCEMENT GAP

If American environmental regulations are so much stricter than Britain's, why has the United States not made substantially more progress in improving the quality of its environment? The explanation is a simple one: American environmental laws and regulations have not been uniformly enforced. Enforcement has been relatively strict in some areas, lax in others. Accordingly, if one turns from an examination of each nation's rules and regulations to the way in which these controls have actually been implemented, the significance of the differences in the severity of British and American regulations, while it by no means disappears, diminishes substantially.

While air and water in the United States are undoubtedly cleaner than they would have been in the absence of the laws enacted fifteen years ago, none of the goals embodied in legislation in the early 1970s has actually been achieved, nor is any likely to be in the foreseeable future. The deadline for the strictest and most ambitious standard—for automobile emissions—has been postponed four times, twice by legislation and twice by the EPA. The most recent revision, included in the Clean Air Act Amendments of 1977, gave the automobile industry an additional five years to meet the original hydrocarbon standard and five to seven years to meet the original standard for carbon monoxide emissions. Compliance with the original nitrogen oxide standard was delayed for an additional five years, and the standard itself was relaxed from 0.4 grams of pollutant to 1.0; under certain circumstances it is allowed to exceed 1.5 grams. As a result, by 1977, when new car emissions were originally supposed to be reduced by 90 percent, they had actually been reduced only 67 percent.

Moreover, even those standards that the automobile manufacturers have in fact met do not necessarily translate into an equivalent reduction in emissions. While the 1970 amendments require that vehicles meet federal emission standards for a minimum of 50,000 miles, according to the EPA's own estimates, none in fact does so. For the most recent model years, the average automobile after 50,000 miles of use emitted three times as many hydrocarbons, nearly four times as much carbon monoxide, and twice as much nitrogen oxide as when it left the factory. Even these figures exaggerate compliance, since most cars are generally driven far beyond 50,000 miles. Not only does the effectiveness of catalytic converters—the devices designed to assist in meeting the original emission standards—diminish with use, but a considerable number of automobile owners have tampered with them in order to improve fuel economy and save the additional expense of unleaded gasoline.[43]

If one turns to the record of industrial performance in the area of air-pollution control, one finds an equally patchy record with respect to compliance. According to a 1976 Senate committee report, six years after the enactment of the 1970 Clean Air Act Amendments, "out of roughly 22,000 major emitting facilities, at least 3,000 either did not comply with emission limitations or did not adhere to compliance schedules."[44] As of 1981, 87 percent of the nation's integrated iron and steel facilities, 19 percent of its other iron and steel factories, 21 percent of its petroleum refineries, and 54 percent of its primary smelters had yet to comply with the emissions limits established by state and federal agencies.[45] According to the Council on Environmental Quality, approximately 1,700 major pollution sources (those with a potential to emit more than 100 tons of pollutants per year) were not controlled adequately in 1981.[46] (Moreover, these figures are based primarily on the companies' own records, which may not always be reliable: according to the EPA, tests of actual emissions account for less than 5 percent of all compliance determinations.) In addition,

> the sources still exceeding their legal limitations include not just the 8 to 10 percent officially listed as in violation, but the equal number of sources— about 1,200—that are meeting schedules that will not bring them into full compliance for many years.... The General Accounting Office estimates that almost 25 percent of the sources regarded by the EPA as in final compliance actually are exceeding their emissions limitations.[47]

The EPA has been no more successful in enforcing the even more ambitious goals established by the 1972 Clean Water Act. In 1977 Congress again amended the Clean Water Act: firms were now given until July 1, 1987, to install the best available technology for nonconventional pollutants. Congress did not, however, alter its original goal of making all lakes and streams in the United States swimmable and fishable by 1983; by that date only half of the nation's lakes and streams had actually achieved that standard. As of 1981, moreover, 22 percent of major municipal treatment facilities and 15 percent of the nation's major industrial plants had yet to comply with the terms of the permits they had been issued nearly a decade earlier.

While it is certainly true that the amount of compliance achieved by many industries is greater than their bitter opposition to technology-forcing standards and strict deadlines would have led one to expect, there nonetheless remain real-world limits to the amount of pollution control that industry can achieve within any given period of time. In some cases the technology to reduce emissions below a specified level did not in fact exist, or its adoption—particularly in the case of older facilities—was so expensive as to render a substantial number of plants unprofitable. In principle there was nothing to prevent the American government from establishing rigorous environmental standards, strictly monitoring emitter behavior, and then either imposing heavy financial penalties on firms not in compliance or refusing to issue permits for the construction of new or expanded facilities that would reduce environmental quality. In practice, however, while a strategy of rigorous enforcement is viable in particular cases, it can hardly be applied across the board. There are limits to the amount of economic disruption the citizens of any democratic nation will tolerate: the law ends precisely when the costs of compliance become excessive.[48]

Noting the inability of one of the major automobile companies to comply with automobile emission standards in the mid-1970s, Helen Ingram observed: "The federal government has the ultimate weapon—shutting down industry—but its use, like that of the A-bomb, was unthinkable.... The tough talk of the Clean Air Act—imposing heavy monetary penalties for failure to meet the standard—was whistling in the dark when the economic consequences were considered."[49] In fact, the EPA has never attempted to shut down major facilities that were otherwise economically viable. According to one study, the capital cost of achieving the goal of zero discharge into the nation's waterways by 1985 would require the allocation of the entire gross national product of the United States for a year; clearly, no agency can be expected to enforce this statute.[50] The Toxic Substance Control Act of 1976 required that industry refrain from employing any chemical substance that posed "an unreasonable risk to health or the environment."[51] The substances used or manufactured by industry that could potentially fall into this category number in the tens of thousands. Given the considerable time and money necessary to test each one, any effort to enforce this statute effectively would disrupt much of the American economy. In fact, it has hardly been enforced at all. A National Academy of Science study reported in 1984 that less than 2 percent of suspected carcinogens had been sufficiently tested to allow their health hazard to be assessed, while for more than 70 percent "there is no information on possible effects on human health."[52]

Moreover, Congress has repeatedly proved itself responsive to requests for "regulatory relief" from particular industries: over the last decade the automobile industry, the smelting industry, shopping centers, high-sulfur coal producers (all in 1977) and the steel industry (in 1981) have been granted some form of relief from the provisions of the 1970 Clean Air Act Amendments.[53] And while a number of energy projects off the Pacific and Atlantic coasts were delayed or prevented by environmental regulations during the 1970s, in the case of the most important domestic energy project—Alaska oil—Congress, like Parliament, promptly approved legislation that specifically bypassed its own environmental procedures following the Arab oil embargo of 1973. And in 1983 Congress prevented the EPA from imposing a construction ban on states that had not met the Clean Air Act's deadline for attaining primary air-quality standards.

Part of the enforcement problem has been administrative in nature. During the first half of the 1970s the EPA found itself overwhelmed by the sheer complexity of enforcing the recently enacted Clean Air and Water Act Amendments, let alone its regulations governing noise, drinking water, hazardous wastes, pesticides, and other toxic substances. The EPA was required, for example, to issue permits for more than 60,000 industrial and municipal discharges into the nation's waterways; four years later it had managed to issue permits for only two-thirds of all industrial discharges. Moreover, many of these permits were issued before the agency had been able to promulgate the effluent limitations required by the 1972 statute and thus were inconsistent with the effluent limitations guidelines that the agency eventually adopted. Thanks to court challenges by industry, it was not until 1977 that the EPA was able to establish an administrative framework for implementing the provisions of the 1972 Water Quality Act requiring all emitters to employ the best practicable technology—by 1977. As of 1980 the EPA had yet to act on 643 proposed changes in state-established controls on existing sources of air pollution. New source performance standards had still not been issued for a number of major air-polluting industries and important industrial sources of water pollution had yet to be given effluent guidelines.

Once regulations were issued, the EPA was faced with the challenge of enforcing them. But while the agency devoted considerable time and legal resources to defend its authority to enforce "infeasible regulations," fearing a political backlash from industry, it made little effort to do so. Instead it began to issue administrative orders that either gave firms additional time to comply or allowed them to "cut a few corners by installing controls of marginal effectiveness."[54] These orders, as one environmental law book put it, represented "a variance mechanism excusing widespread non-compliance."[55] In the United States, as in Britain, newly constructed plants have been required to meet stricter pollution-control requirements than existing facilities. But again, with a few highly publicized exceptions, these regulations have also been interpreted relatively flexibly. According to a study published by the Conservation Foundation, "it is . . . becoming apparent that, although federal laws are essentially uniform as written, techniques of monitoring and calculating pollution impacts—both critical components in the review of environmental permit applications—are subject to a fair amount of finagling and negotiation."[56]

Moreover, even when the EPA has tried to demand strict compliance from industry, it has often been unable to achieve it. Indeed, in the case of air pollution, it is precisely the most heavily polluting industries that have been most successful in using both the courts and the Congress to delay installing adequate abatement technology. Approximately twenty-five lead, copper, and zinc smelters located in western states account for one-tenth of the nation's sulfur dioxide emissions: two of these smelters account for more than one-third of the sulfur dioxide emitted west of the continental divide. Yet fifteen years after the passage of the 1970 Clean Air Act Amendments, most of these facilities had yet to establish compliance schedules that would enable the regions in which they are located to meet national ambient air standards. Approximately one-third of the sulfur dioxide emitted east of the Mississippi comes from coal-burning power plants in three states: Ohio, Indiana, and Illinois. Yet "midwestern utilities have proven nearly as successful in combining litigation with appeals to state governments and Congress as have smelters," and the EPA has been unable to secure the adoption of enforceable state implementation plans that would restrict sulfur dioxide emissions in the Ohio River Valley.[57] According to the Government Accounting Office, in one EPA region only half of the 321 major air polluters not in compliance had ever had any enforcement action taken against them.

There have certainly been cases in which the enforcement of environmental regulations has been stricter in the United States than in

Britain. Clearly the American automobile industry has been forced to meet stricter pollution-control requirements than its British counterpart. Oil companies have found it easier to expand their refining capacity in Scotland than on the east coast of the United States. American chemical producers have experienced stricter controls on the manufacture and distribution of toxic substances than firms operating in Britain, and both nuclear and nonnuclear energy projects in the United States have been forced to meet stricter safety standards than those constructed in Britain. And while approximately 10 percent of utilities in the United States have installed scrubbers, British power plants have been permitted to disperse their emissions by constricting tall stacks—a far less costly means of abatement.

On the whole, though, it does not appear that the balance struck between amenity and economic values has varied significantly in the two countries: the Americans have enacted more rigorous and comprehensive regulations but have experienced much greater difficulty in enforcing them. In spite of the intent of Congress to establish uniform standards, industrial compliance with them has been no less subject to bargaining, negotiation, and compromise than has been the case in Britain.

In sum, the relatively high degree of acceptance of environmental regulation on the part of the British business community cannot be attributed to the "capture" of the former by the latter. On balance, environmental policy has been no more or less influenced by economic considerations in America than in Britain.

THE COSTS OF COMPLIANCE

The validity of this conclusion is suggested by data on the costs of pollution control in the two countries. According to figures collected by the Department of Commerce, total pollution-control expenditures averaged slightly more than 1.8 percent of GNP between 1975 and 1980, declining to 1.7 percent of GNP in 1981 and 1.65 percent in 1982.[58] In 1981, when the Department of the Environment published its estimates of British pollution-control costs for 1977–78, it reported that total expenditures on all forms of pollution control came to slightly less than £2.5 billion, or between 1.5 and 2 percent of GNP at factor cost. Admittedly this comparison is a crude one. Even if we assume that the bases for calculating pollution-control expenditures are comparable in the two countries, such expenditures represent only a portion of the costs of environmental regulation. They omit, for example, the opportunity costs of potentially profitable investments that have been delayed or with-drawn because of environmental controls and the costs associated with the diversion of research and development into compliance with gov-ernment regulations. On the other hand, they also take no account of the economic benefits from expenditures on environmental improve-ments, which appear to be both considerable and difficult to measure in both countries.

For our purposes, what is critical is not so much the shares of national resources devoted to environmental protection, which, at least according to the official statistics of the two governments, appear to be roughly equivalent, but rather the extent to which these expenditures have been borne by each nation's business community. Have environmental ex-penditures in fact imposed a greater economic burden on American than on British business, and might this perhaps explain why environ-mental regulation has created so much more political conflict between industry and government in the United States?

It is difficult to answer this question satisfactorily. In Great Britain approximately half of the pollution-control expenditures were made by the private sector; in America the comparable figure was two-thirds. Much of this difference, however, might be explained by the larger size of the public sector in Great Britain. On the other hand, pollution-control expenditures by private industry qualify for special tax treatment in the United States—they can be depreciated 100 percent over five years—but not in Great Britain.[59] (This difference may also serve to exaggerate the costs of industrial compliance in the United States.) Corporations in America can also finance expenditures on pollution control by issuing special pollution-control bonds, which carry a lower interest rate than ordinary taxable bonds; most of the steel industry's expenditures on pollution control have been financed in this manner. As a result, a sig-nificant share of private-sector expenditures in the United States has actually been financed by taxpayers.

We also have no basis for determining the extent to which a particular industry was able to pass on the costs of compliance to consumers or for measuring the aggregate impact of environmental expenditures on each nation's rate of productivity growth, inflation, employment, and balance of payments. Unfortunately, while we have numerous studies on each of these dimensions for the United States—most of which suggest that the impact of environmental controls on the overall performance of the economy has not been significant—we have none for Britain.[60] Finally, the time frame must be taken into consideration. For while in all like-lihood American industry has allocated more resources to environmental protection during the 1970s, were we to compare American and British expenditures over a twenty-five-year period, the magnitude of these differences would probably diminish considerably.[61]

My conclusion, admittedly based on incomplete evidence, is that environmental regulation has not been an important cause of the economic difficulties confronted by either nation over the last decade. Moreover, even if compliance with environmental regulations has imposed a somewhat greater burden on industry in the United States than in Britain, the differences do not appear to be of sufficient magnitude to account for the contrasts in the political responses of industry to environmental regulation in the two countries. They cannot explain why the British business community—which includes both nationalized and privately owned firms—is so much more satisfied with its government's system of environmental controls than its counterpart in the United States. Why are executives in America much more likely to attribute their economic difficulties to environmental regulation than their counterparts in Great Britain? Why, even though, on balance, the enforcement of environmental controls has been no stricter in the United States than in Britain, has environmental regulation created so much more political conflict between business and government in the United States?

## POLITICAL IMPLICATIONS

While it would be misleading to dismiss entirely the importance of the relative costs of compliance in accounting for the different political responses of business in the United States and Great Britain to environmental regulation over the last fifteen years, their role does not appear to be decisive. Both critics and defenders of environmental policy in the United States have tended to focus on the economics of compliance—the former to document how burdensome the costs have been and the latter to demonstrate their relative unimportance. Both perspectives miss the point: the distinctively adversarial nature of environmental regulation in the United States has less to do with economics than with politics. The system is not an adversarial one because the costs of compliance are so high; rather it is the adversarial nature of American environmental regulation that makes both the direct and indirect costs of compliance appear excessive. It is the way in which environmental policy is made and implemented, not the direct cost of complying with it, that accounts for the resentment it has aroused within the American business community and its relatively high degree of acceptance on the part of business executives in Britain.

## Participation

One critical difference in the ways environmental policy is made and enforced in Britain and in the United States has to do with the role of business. In Great Britain, business participation in the making and implementation of environmental policy is both assumed and assured. Compared to that of other political constituencies, most notably that of environmentalists, the political position of business is clearly a privileged one: it is closely consulted before pollution controls are both made and enforced, and developers alone enjoy the automatic right to appeal the decisions of local planning authorities. While business does not always win, its views are always given careful consideration by government officials and its access to policy makers is assured by both law and custom.

In America, on the other hand, while business certainly does not lack opportunities to influence environmental policy, its participation is neither assumed nor assured: it must constantly be asserted. Thanks to the liberalized rules governing access to the federal courts, business enjoys few legal privileges not available to environmental groups. And its political resources are certainly far more closely matched by environmental organizations than is true in Great Britain. While the ultimate outcomes of environmental policy in the two societies may not vary substantially, in Great Britain the balancing of economic and environmental considerations is built into the policy process. In America, however, the importance given to economic considerations is in large measure dependent on the lobbying and litigation skills of business.

As a result, American executives have been forced to devote far more political and legal resources to efforts to influence environmental policy than their counterparts in Great Britain. Senior British managers rarely concern themselves with environmental issues. While they frequently consult with government officials, these discussions almost never focus on matters of environmental policy; such matters are discussed at a much lower level, generally by technical personnel in both sectors. In many American industries, however, environmental policy has been a major preoccupation of senior corporate managers over the last fifteen years. Many chief executive officers in the United States have been actively involved in shaping the regulatory strategies of their companies and have on occasion personally lobbied members of Congress and the executive branch. The Business Roundtable, an organization of the chief executive officers of most of the nation's largest corporations, has devoted a substantial share of its resources to attempting to influence environmental policy, particularly in the area of air pollution.

While both the American and British business communities became much more politically active during the 1970s, only in America can such activities be attributed to the increases in government regulation in general and environmental regulation in particular. In Britain they have represented primarily responses to pressures from trade unions and the growth in the size of the public sector.[62] The major organization of British business, the Confederation of British Industry, has become somewhat more involved in environmental issues since the mid-1960s: a number of senior staff members are responsible for preparing and presenting the position of CBI on proposed environmental policies to officials in Whitehall, and an Environmental and Technological Legislation Committee regularly monitors and attempts to influence private members' bills that are likely to affect the interests of industry. These matters, however, constitute a relatively minor part of the CBI's activities. Some individual companies have been forced to play a more active and aggressive role in persuading local planning authorities to grant planning approval, but they are relatively few. While the amount of attention the staffs of British trade associations have devoted to environmental issues has certainly increased, neither the scope nor the nature of their interaction with government has undergone any substantial change.

The contrast with the United States is dramatic. Corporate legal expenses and public and governmental relations budgets have all substantially increased over the last decade, in large measure as a response to environmental regulation. Whereas in Great Britain, lawyers are involved only in local public and parliamentary inquiries, the legalistic nature of American environmental regulation requires the involvement of lawyers at every stage of the regulatory process: they represent companies both in court and in administrative proceedings. And to a far greater extent than in Great Britain, both corporations and trade associations in the United States have committed substantial resources to public relations in an effort to persuade the public of the degree of their commitment to environmental quality and of the merits of their positions on particular issues. The greater complexity of the American regulatory process also has made the administrative costs of compliance substantially greater for firms in the United States.

Expenditures for corporate lobbying have also increased substantially: the number of corporations with offices in Washington increased from 100 in 1968 to more than 500 in 1978. Such trade associations as the American Petroleum Institute, the Lead Industries Association, and the American Motor Vehicles Manufacturing Association have waged lengthy legislative and legal battles challenging particular environmental standards, while numerous new industrial associations, such as the Utilities Air Regulatory Group, the National Economic Development Association, and the American Industrial Health Council, have been established for the express purpose of influencing environmental policy at the federal level.

Underlying these developments was a more substantial shift in the balance of power between industry and environmental groups in the United States. The political influence of environmental organizations in Britain did increase during the late 1960s and early 1970s, but they challenged rather than threatened the close ties that existed between British industry and government with respect to the making and enforcement of environmental policy. Their influence in Whitehall, while it increased, never approached that of industry, and Britain's constitutional system limits Parliament's ability to challenge government decisions. Environmental organizations have no standing in the British judicial system to challenge administrative decisions and the British government has refused to defray any of the expenses they incur in order to present evidence at either local or large public inquiries. Only in the areas of land-use planning and transport policy has the ability of nonindustry constituencies to participate in the policy process notably increased since the late 1960s. On balance, probably the most important and lasting increase in the power of amenity groups in Britain over the last fifteen years has been at the local level, particularly in the more affluent communities in the south of England.

In America the situation has been much different. Industry, which had enjoyed an effective veto power over environmental legislation in the postwar period, began to experience an erosion of its influence in Congress in the late 1960s. By the early 1970s, the environmental movement had emerged as a powerful political lobby. Compared to the political influence of various constituent groups of the public interest movement, that of industry remained relatively stable over the next decade. During the second half of the 1970s, environmental groups were able, with a few exceptions, to resist industry's efforts to weaken the laws enacted during the early 1970s; indeed, in a number of cases, through judicial intervention, they were able to strengthen them. Environmental activists occupied important policy-making positions in the Carter administration, which held office between 1977 and 1980. Even under the first Reagan administration, the environmental movement was remarkably successful in resisting industry's effort to rewrite the nation's environmental laws and weaken their enforcement. In no other policy area did the administration's efforts to provide "regulatory relief" to industry meet with such effective opposition, which contributed to the resignation of both the secretary of the interior and a number of senior EPA officials.

The nature of the environmental movement itself also changed more substantially in the United States than in Great Britain.[63] While the membership of environmental organizations increased in both societies, those organizations in Britain that reported the greatest increases in membership—the National Trust and the Royal Society for the Protection of Birds—are only peripherally involved in political activity; they function more as philanthropies and administrative bodies than as pressure groups. (Many people join them because of the benefits they provide, which in the case of the National Trust include a discount on the price of admission to stately homes registered with the Trust.) Environmental organizations in Britain continue to devote a major portion of their energies and resources to assisting civil servants in implementing public policies. Moreover, many of the political efforts of British environmental organizations have been confined to the local level; a disproportionate share of their energies has focused on issues of immediate concern to their members, such as a planning decision or the pollution from a local plant.

The American environmental movement has become much more politicized. For virtually every environmental organization in the United States, influencing public policy has become its most important priority. This is true not only of the newer organizations, such as Friends of the Earth and Environmental Action, but also of older, established ones, such as the Audubon Society and the Sierra Club. The Audubon Society does manage 76 sanctuaries and 90 preserves and the Nature Conservancy has preserved approximately two million acres of land.[64] But such activities take up a much smaller part of environmentalists' energies in the United States than in Britain. In addition, the American environmental movement has consistently demonstrated greater ability to wage national campaigns. Finally, American environmental organizations employ larger staffs and pay them higher salaries than their counterparts in Britain.

Over the last fifteen years American environmental organizations have acquired extensive rights to information about various public policies, the right to challenge a wide variety of administrative decisions in the courts, and the right to be heard in various administrative proceedings. British environmental organizations, by contrast, possess remarkably few rights. The information they receive from government officials, while it has increased, remains extremely limited: Britain has nothing even remotely resembling the Freedom of Information Act. Unlike American environmental groups, their British counterparts have no right to be consulted by government officials; whether or not they receive consultation papers is up to the discretion of ministers and senior civil servants.

Only in the case of land-use decisions do British environmental groups enjoy anything resembling administrative due process, and even here the government has limited the standing of organizations not directly affected by particular decisions.

Environmental organizations in both countries have received financial support from their respective governments. In Britain such support has frequently involved either direct grants or the delegation of various governmental functions to nongovernmental bodies. In America it has been more indirect.[65] Under the "private attorney general" concept, plaintiffs in successful court actions can, under certain circumstances, have their court costs as well as other expenses reimbursed. Since 1974, public-interest law firms have been able to sue government agencies and private organizations without forfeiting their tax-exempt status; public-interest groups also can devote a certain portion of their resources to lobbying without losing their tax exemptions. In addition, they are able to solicit contributions and membership dues through the mail at special bulk rates available to nonprofit organizations.

What is critical is not the relative magnitude of this assistance in the two countries but its political implications. In America, not only do they provide little incentive for environmental organizations to cooperate with government officials, but they have actually facilitated their ability to challenge both industry and government.[66] In contrast, the direct and indirect financial assistance provided by the British government is contingent on their behaving responsibly; they serve to reduce rather than exacerbate the adversary process. The control of British officials over access and their ability to determine which groups are invited to have their representatives serve on advisory committees or receive consultation papers increase the dependence of environmental organizations on civil servants and ministers. The prominent role played by the courts in American environmental policy has precisely the opposite effect: it enhances their ability to challenge administrative decision.

### The Pace of Change

During the postwar period the scope of environmental regulation increased, its administration became more centralized, and opportunities for public participation were expanded in both countries. But what is striking in the United States is not only the relative magnitude of these changes but also their compression into a relatively short time period: in the space of three years—between 1969 and 1972—the nature and administration of environmental regulation changed more substantially than it did in Great Britain during the preceding quarter century.

While the British central government steadily expanded the scope of environmental regulation in the postwar period, its basic approach to environmental regulation has remained remarkably stable. The authority of the Alkali Inspectorate was substantially increased, but it continued to operate under the statutory framework established by the Alkali, Etc. Works Regulation Act of 1906. (This legislation was formally superseded by the Control of Pollution Act in 1974 but its substance remained unchanged.) The establishment of the regional water authorities in 1972 constituted an important administration reorganization, but it did nothing to alter the way in which government officials went about controlling industrial emissions. Only with respect to smoke emissions did public policy change significantly, but the control of smoke emissions remained, as before, in the hands of local officials: the Clean Air Act (1956) essentially expanded their authority. And in spite of the increased public concern with toxic substances, the control of pesticides has continued to be governed by the system of industry self-regulation established in 1958.

The establishment of the Department of the Environment, for all it signified about the government's commitment to environmental protection, represented little more than an administrative reorganization: the department was granted neither new authority nor additional regulatory responsibilities. In a sense its title is misleading; only a small portion of its responsibilities have to do with environmental regulation. Its only new component, the Central Unit on Environmental Pollution, was a coordinating body. Equally important, the nature of the staffing of the various British regulatory bodies underwent no change at all: regulations continued to be written and implemented primarily by technically and scientifically trained personnel. Parliament continued to delegate broad regulatory authority to ministers, who in turn left virtually all important decisions to civil servants: parliamentary oversight has remained modest. The absence of statutory deadlines, a reluctance to prosecute, the emphasis on cooperating with industry rather than coercing it, and a flexible and decentralized approach to the making and enforcement of rules—all have remained consistent features of the British approach to pollution control for nearly a century.

By contrast, the strategy of the American government for controlling pollution changed substantially. The most important shift was associated with the transfer of authority from the states to the federal government. The federal government, for the first time, established both national ambient air standards and uniform emission standards for new plants (new source performance standards). The primary responsibility for controlling water pollution was also transferred from the states to the federal government. In addition, the amount of administrative discretion was substantially reduced; the laws enacted in the early 1970s not only included specific deadlines but presented agency officials with relatively strict guidelines for both emission standards and new pollution-control technologies—requirements whose enforcement could then be challenged in the courts.

While the federal government's previous controls over water pollution were based primarily on water-quality standards, the 1972 Clean Water Act Amendments abandoned this approach in favor of technologically determined emission standards. The Toxic Substances Control Act both transferred regulatory authority from the Department of Agriculture to the EPA and required the reregistration of all previously approved substances. In addition, the National Environmental Policy Act of 1969 required that every government agency prepare an extensive "environmental impact statement" before approving any new development—a requirement that had never before existed in the United States.

The establishment of the Environmental Protection Agency, unlike that of the Department of the Environment, significantly changed the dynamics of government regulation. Not only did it make it more centralized, but it contributed to a decline in the influence of scientists and engineers in the making and enforcement of regulatory policy. Most of the staffs of administrative units that were merged into the EPA had been primarily technical in their orientation and training. Within the EPA itself, however, under the leadership of its first administrator, William Ruckelshaus, who had previously worked in the Department of Justice, lawyers rapidly assumed a more important role in the nation's pollution-control efforts. Promising to enforce the nation's pollution-control laws vigorously, "Ruckelshaus stressed EPA's enforcement duties as opposed to its research responsibilities."[67] The agency brought five times as many enforcement actions during its first two months as all its predecessor bodies had initiated in any comparable time period. Reviving the long-dormant 1899 Refuse Act, EPA lawyers brought 371 enforcement actions to the Justice Department for prosecution—and sought criminal penalties in 169 of them.

Both the rapidity and the magnitude of these changes clearly created a considerable strain on American business: executives suddenly found themselves forced to adjust to a dramatically changed regulatory environment. Unlike their counterparts in Great Britain, who continued to deal with essentially the same civil servants enforcing the same regulations in much the same manner, American executives were confronted not only with a whole new series of laws and regulations but with a new regulatory bureaucracy with whose personnel and procedures they were unfamiliar. While the number of officials responsible for environmental regulation has remained relatively stable in Britain over the last decade, the staffing of environmental regulatory agencies at both state and national levels in the United States increased significantly.

Moreover, the uncertainty created by the dramatic shifts in environmental policy in the late 1960s and early 1970s did not appreciably diminish for the remainder of the decade. On the contrary, no sooner were these laws enacted than a prolonged battle broke out over their interpretation, first within the EPA and subsequently within the federal courts. Within a relatively short period of time, both environmentalists and industry pressed for new legislation. The battle over the amendments to both the Clean Air and Clean Water acts preoccupied industry throughout the greater part of the 1970s; indeed, the Clean Air Act Amendments of 1977 were among the most intensively lobbied laws of the postwar period. Thus important elements of American environmental policy have remained in a continual state of flux; executives have found themselves constantly vulnerable to significant and often abrupt changes in public policy, whether initiated by Congress, the executive branch, or the courts.

Environmental policy has also been much more affected by the electoral process in America than in Great Britain. Enforcement efforts were significantly strengthened following the election of Jimmy Carter in 1976, in part because the administration appointed environmental activists to important policy-making positions. The Clean Air Act Amendments of 1977 measurably tightened controls over stationary source emissions, and in 1980 Congress enacted the Comprehensive Emergency Response, Compensation, and Liability Act (Superfund). British environmental policy was much less affected by the election either of a Labour government in 1974 or of a Conservative government in 1979. Unlike the Reagan administration, which undertook some highly controversial initiatives designed to weaken the enforcement of environmental legislation in the United States, the Thatcher government has not attempted any major change in British environmental policy: on balance it has neither strengthened or weakened controls over industry. It did permit an increase in the weight of heavy lorries and has attempted to expedite the approval of planning applications, but it also moved to phase out the use of lead in petrol and indicated in 1984 that it would strengthen planning controls over agricultural land use.

*The Enforcement Gap*

Another set of factors has served to exacerbate tension between industry and government in the United States: the enforcement gap itself.

While the balance struck between economic and amenity values in the United States may not differ significantly from that struck in Great Britain, this is true only in the aggregate. But companies do not live in the aggregate; what concerns them is the way the particular regulations that affect them are enforced. An important consequence of enacting laws that cannot be enforced has been the creation of considerable uncertainty on the part of those who may have to comply with them: individual companies and even entire industries can never be sure if EPA or state officials will decide to enforce the letter of the law in their particular case. Moreover, because enforcement is not—indeed, cannot—be uniform, when it does in fact take place, companies are likely to feel that they have been singled out unfairly: enforcement invariably appears arbitrary. Moreover, companies that have made a good-faith effort to comply with environmental regulations are likely to become resentful of those who have not, particularly when the latter's lack of compliance reflects not so much their objective economic difficulties as their political clout.

The fact that many American environmental regulations have been written in such a way that they cannot be enforced means that government officials, unlike those in Britain, receive no credit for enforcing them in a flexible manner. They are viewed as having made concessions rather than compromises. While British industrialists attribute their government's system of flexible enforcement to the "good sense" of their nation's officials, the American business community tends to attribute the "regulatory relief" granted them by either Congress, regulatory agencies, or the courts as a testimony to their own lobbying and litigation skills. Rather than getting credit for being "reasonable," the government is blamed for enacting "unrealistic" regulations in the first place. Likewise, the efforts of American industry to require that the government conduct a cost-benefit analysis before issuing new regulations reflect business's perception that the costs of compliance are not carefully weighed before regulations are promulgated. Their counterparts in Britain have rarely had reason to doubt that regulatory requirements will be tailored to what they can afford to spend. As a result, the economics of enforcement may be similar in both nations, but the political and legal context in which they take place differs markedly.

When regulations have been strictly enforced in the United States, the enforcement has often been done in a manner that appears unreasonable.[68] Studies of environmental regulations published over the last decade contain numerous accounts of companies that were forced to install abatement technologies that had not yet been adequately tested, of fines that were levied for unintentional violations of various regulations, of companies that were forbidden to employ the most cost-effective means of pollution control by the rigid application of particular rules, and of the enforcement of rules that actually reduced environmental quality.[69] The point is not that the strict and literal enforcement of environmental regulations has posed an intolerable burden on American industry; that is clearly not the case. Nor is it relevant that the examples of "unreasonable enforcement" so frequently cited by critics of environmental regulation may not be typical of the way in which environmental regulations are actually enforced; that is also beside the point. What is important is that in the United States, but not in Britain, such incidents have occurred with sufficient frequency to undermine the legitimacy of environmental regulation in the eyes of the companies that have to comply with it.

### The Politics of Risk

There is another reason why environmental regulations appear more reasonable to business executives in Britain than in the United States: in Britain they are based on a consensus among scientists and engineers in both business and government. In the area of pesticide regulation, for example,

> rather than publicly confronting each other, toxicologists have been enlisted by the British government to generate a consensus and legitimate political decision. In contrast to the conflicts among experts that characterize many American decisions in this field, British decisions emerge from a closed decision-making process with the apparently uncontroversial and authoritative support of science, where U.S. decision-making institutions depend upon and, to some extent, generate conflicts among experts.[70]

While industry and government scientists in the United States have often disagreed bitterly as to whether or not particular substances constitute an appropriate "margin of safety," in Great Britain such judgments rarely divide officials in the two sectors. In general, while both nations have acted in a roughly similar fashion to regulate products, production processes, and pollutants whose harm to public health has been clearly demonstrated, American regulatory agencies have been far more willing to impose regulations on the basis of inconclusive or fragmentary scientific evidence in order to provide an "adequate margin of safety." British regulatory policy has tended to confine its attention to known hazards, while American officials have been far more active in seeking to anticipate—and thus avoid—future harms. British regulatory policy has been relatively cautious and incremental in its orientation; officials tend to study a problem carefully before making a policy decision. In America, on the other hand, in part because of congressional pressure, officials have made regulations relatively hastily, often before they have had the opportunity to fully assess the scientific evidence on which they were purportedly based.

The characteristic American policy toward risk can be seen in a 1976 Court of Appeals decision upholding the EPA's ambient air standard for lead. The court reasoned:

> Petitioners [i.e., the industries] argued that the "will endanger" standard requires a high quantum of factual proof, proof of actual harm rather than of a 'significant risk of harm'.... We have considered these arguments with care and find them to be without merit.
>
> ... A statute allowing for regulation in the face of danger is, necessarily, a precautionary statute. Regulatory action may be taken before the threatened harm occurs; indeed, the very existence of such precautionary legislation would seem to demand that regulatory action proceed, and, optimally, prevent the perceived threat.... The statutes and common sense *demand* regulatory action to prevent harm, even if the regulator is less than certain that harm is otherwise inevitable.[71]

A British social scientist writes that "Americans seem to have taken an excessively strict interpretation of risk, reducing 'reasonable risk' practically to zero risk." For example,

> the primary standard for photochemical oxidants is based on ozone concentrations (itself a questionable surrogate substance) where the 'zero risk' level for one hour, once per year is 0.08 p.p.m. (compared with a primary annual standard of 0.365 p.p.m.).... But, there appears to be no sound scientific evidence that such levels are 'requisite to protect the public health' (as defined in the Act). It is quite possible that no *unreasonable* public health risk occurs with concentrations three times this level, so it is hardly surprising that the U.S. Department of Commerce, backed by American business, is trying to persuade Congress to amend the working of the Act.[72]

The Clean Air Act (1956) represented a response—albeit a rather belated one—to the clear dangers to public health produced by the domestic burning of coal. By contrast, the far more stringent controls over automobile emissions enacted by the Clean Air Act Amendments of 1970 were approved without any scientific evidence as to the threats to public health posed by the pollutants whose reductions they mandated.

While these reductions have undoubtedly improved the health of segments of the population, to date "there have been no efforts at direct measurement of the health improvements or other ultimate benefits that may actually have occurred as a consequence of the mobile source program"; indeed, a 1977 study revealed no correlation between public health and the concentration levels of the three principal pollutants emitted by automobiles.[73]

An article in *Social Studies of Science* contrasts the approaches of British and American authorities to the regulation of aldrin and dieldrin, two closely related organic pesticides widely used in agriculture in both countries during the 1960s. Laboratory studies revealed that the pesticides increased the incidence of liver tumors in mice, but did not induce cancer in any other organs of this animal and failed to produce a carcinogenic response in rats, monkeys, or dogs. The British government concluded that there was insufficient evidence that these pesticides caused malignant tumors in humans; the British "expected the traditional requirements of scientific causality to be satisfied before labelling a chemical carcinogenic." The EPA, on the basis of exactly the same scientific data, banned the pesticides after interpreting the Federal Environmental Pesticide Control Act to hold that "suspension is to be based upon potential or likely injury and need not be based upon demonstrable injury or certainty of future public harm."[74]

On several occasions over the last fifteen years the United States government has taken prompt regulatory action only to have the scientific rationale on which it was based subsequently undermined. In 1980, for example, President Carter evacuated 700 families from their homes near Love Canal in New York State on the basis of a study commissioned by the EPA that reported an unusually high incidence of chromosome abnormalities; a subsequent, more thorough study by the Center for Disease Control reported that when the families were compared to a control group, no excess abnormalities were found. While DDT was banned in 1972 because it caused liver tumors in several dozen mice, a subsequent two-year study of DDT in both rats and mice by the National Cancer Institute found "no evidence for the carcinogenicity of DDT in rats and mice." In the mid-1970s, responding to the arguments of environmentalists that synthetic chemicals in aerosol spray cans were destroying the atmosphere's protective layer of ozone and thus increasing the likelihood of deaths from skin cancer, the EPA, along with other government regulatory agencies, banned most deodorants and hairsprays. This move forced the alteration of more than 30,000 consumer products. Yet in 1983 the National Academy of Sciences reported that there was "no discernible change in the ozone level between 1970 and 1980 and that it is possible that other industrial by-products such as car exhausts may even be increasing the ozone level."[75]

Faced with similar kinds of evidence, the British have acted more cautiously. They have imposed fewer restrictions on the use of DDT, for example. They have tightened standards on TCDD, a toxic dioxin that is a component of the herbicide 2,4,5-T, but have not chosen to ban its use; indeed, officials at the Ministry of Agriculture contend that "the American environmental agency was panicked into its partial ban by a now discredited study in Oregon which linked spontaneous abortions to 2,4,5-T."[76] Compared to the quality of the report used by the Carter administration to evacuate 700 families from Love Canal, the study of the Independent Advisory Group, *Investigation of the Possible Increased Incidence of Cancer in West Cumbria* (1984), is a model of scientific objectivity. Nor have restrictions on aerosol cans been anywhere near as severe in Britain as in the United States. A CBI working paper notes that while British industry was able to wait until a substitute for "hard" detergents was carefully tested before it was adopted, the Americans banned hard detergents only to find that the substitute was far more harmful.[77] The contrast in the regulatory styles of the two nations was eloquently summarized by the British journalist Stanley Johnson in 1971:

> We saw the American thrashing around from one pollution scare to the next, and we were mildly amused. One moment it was cyclamates, mercury the next, then ozone, lead, cadmium—over there they seemed set on working their way in a random manner through the whole periodic table. Over here, of course, we ordered things differently. We took a careful pragmatic typically British approach. Like good jurists, we preferred to believe in innocence till guilt had been proved. We wanted to look at the evidence. How much lead was in the atmosphere? How much lead, if any, occurred naturally in the atmosphere? What relationship was there between lead in the atmosphere and motor-vehicle exhausts? What was the relationship of lead in the human body to exposure to the atmosphere? What was the evidence, anyway, that lead in the blood-stream was harmful to health? Men had died from time to time and worms had eaten them. But had they died of lead?[78]

The British have, of course, since moved to ban the use of lead in petrol, and this policy has been the subject of considerable controversy among both scientists and the general public. But unlike the United States, the British government waited until a considerable body of scientific evidence had accumulated: significantly, the immediate factor precipitating the government's change of policy in 1983 was the recommendation of the Royal Commission on Environmental Pollution, whose members included some of Britain's most eminent scientists. The British government has objected strongly to the EEC Commission's efforts to introduce environmental standards on the grounds that "there is not yet a sufficient body of scientific knowledge on the effects of some pollutants to specify desirable concentration levels in a statutory form." In responding to the Community's draft directive on freshwater standards for fish, for example, the National Water Council pointed out that "fish thrive in 90 percent of British rivers, yet only 50 percent of these waters comply with the Commission's mandatory criteria." (Lord Ashby subsequently remarked in the House of Lords: "Let the freshwater fish and shellfish decide what is good for them.")[79]

The point is not that the tighter restrictions imposed by regulatory agencies in the United States have invariably been ill advised. British officials have on occasion doubtless been too cautious in assessing risks and their citizens may have been inadequately protected as a result. If the Americans acted with excessive haste in enacting the 1970 Clean Air Act Amendments, the British certainly moved far too slowly: nearly four years passed between the killer fog of December 1952 and the passage of the Clean Air Act (1956); moreover, the legislation has yet to be fully implemented. Similarly, notwithstanding the conclusions of the Black report, the British may be insisting on an excessively high standard of proof before moving to reduce emissions from the nuclear processing facility in Cumbria. Similar questions can be raised in regard to the government's official response to allegations that the burning of PCBs in chemical plants at Bonnybridge in Scotland and Pontypool in Wales had led to an increase in birth defects. On the other hand, Britain has not always acted more slowly: it did ban the use of EDB as a pesticide in grain storage facilities two years before the United States acted. What is critical is that while scientific assessments on which officials in Britain base their decisions are almost invariably respected and accepted by the industries affected by them, in the United States scientists from industry and government often share widely divergent views about what constitutes a hazard. The fact that important regulations have been issued in the United States on the basis of studies that subsequently have proved invalid has only served to confirm the skepticism of many corporate executives and scientists about the technical and scientific competence of American regulatory officials.[80]

Not surprisingly, in sharp contrast to the United States, where much of the bargaining over the implementation of environmental policy takes place among lawyers, in Britain such negotiation takes place primarily among technically trained personnel: they interact as experts seeking to devise solutions for particular problems rather than as adversaries attempting to maximize compliance and minimize costs. In their study of "regulatory unreasonableness" Eugene Bardach and Robert Kagen cite an instance that illustrates the importance of this dynamic—or its absence:

> Environmental engineers from regulated firms in California, however much they complained about costs of compliance, expressed much greater satisfaction with regulations formulated by the Bay Area Air Pollution Control District, which has an ongoing technical advisory committee with representations from business and environmentalists and public health experts, than with regulations of the state Air Resource Control Board, which abolished informal advisory committees. Among other reasons, the state board's regulations were constantly criticized for their *technical* deficiencies, for being based on inadequate or old data, or on inadequately tested theories and hence ill-adapted to the actual variety and complexity of production processes.[81]

In general policy makers in Britain have relied far more on the technical and scientific expertise and experience of industry than have their counterparts in the United States. For example, while the Alkali Inspectorate waited until the technology associated with scrubbers had become well developed before requiring their installation in new power plants, the EPA required their installation in new smelters and power plants before many of the technical problems associated with their use had been worked out, with resultant frequent breakdowns and considerable resentment on the part of corporate engineers. Bardach and Kagen report a characteristic incident:

> The state pollution control agency called upon [a] company to install a certain type of scrubber system. The company's environmental engineer complained, "We argued that a scrubber was impractical. I called TVA [which has extensive pollution-control systems]. They were anti-scrubber because of enormous maintenance and breakdown problems. But the agency people said it *could* be done. They *felt they knew better than us!*"[82]

Finally, another important reason why many American pollution-control regulations appear unreasonable is that they frequently have been dictated by considerations other than improvement of environmental quality. In particular, both the regulations designed to prevent significant deterioration and the mandatory scrubbing requirement for coal-burning utilities were motivated in part by a concern to protect the economic and industrial base of the Northeast and discourage companies

from relocating to or purchasing raw materials from the Sunbelt.[83] Whatever the merits of these particular policies, companies have been forced
to incur substantial additional expenses that contribute only marginally
to environmental quality. Great Britain has been no less protective of
the interests of its declining industries and depressed regions; indeed,
if anything, it has been far more so than the United States. But it has
not used environmental policy as a vehicle for achieving this objective.

Expenditures for control of water pollution provide a second illustration of the use of environmental regulations in America as a means for
particular firms to gain a competitive advantage. Since all but a small
portion of the costs of constructing municipal water-treatment facilities
in the United States are financed by the federal government, state and
local governments have had little incentive to scrutinize the costs of
abatement. As a result, the cost of constructing these facilities was inflated
by pressure from construction companies and unions, which regarded
the establishment of strict standards of pollution control as a way of
increasing their income. Consequently, a share of the nation's investment
in water-pollution abatement has produced rather marginal environmental benefits.

Britain, too, has a water-pollution control "industry." Even though it
is public rather than private, its managers and workers also have a stake
in increasing expenditures on abatement. But because the regional water
authorities are meant to be self-financing, their resources are limited.
Since they are organized as businesses—albeit nationalized ones—the
limits placed on their borrowing give them every incentive to use the
limited resources they have available for abatement as efficiently as
possible.

It is difficult to judge which nation's system of pollution control is
more efficient in an economic sense. There is no question, however, that
the inefficiencies of the American system of environmental regulation
are more politically salient. While students of British environmental policy have accused both industry and government of devoting too few
resources to pollution control and failing to establish strict enough controls over potential or actual health hazards, what they have not done
is to challenge the appropriateness of those expenditures that have been
required or those regulations that have been enforced; such criticisms
of the environmental regulations are almost completely absent from
public discussion in Britain. With the exception of some of the reports
of the Royal Commission on Environmental Pollution, there have been
relatively few proposals to improve the efficiency of British environmental policy. In America, by contrast, proposals for reforming Amer-

ican environmental policy have become virtually as numerous as the regulations themselves.[84] While neither nation has made a serious effort to introduce economic incentives as a means of improving the efficiency of its controls over pollution, only in America has there appeared an extensive literature urging policy makers to adopt such an approach.[85]

## Regulatory Complexity

The very nature of American regulations helps explain why environmental policy has produced more conflict between industry and government in the United States than in Britain. The legislative history of the Clean Air Act Amendments of 1977 consists of eight volumes totaling more than 7,500 pages. Christopher Wood writes: "To say that the United States provisions relating to the control of air pollution from stationary sources are arcane would be too generous. They are so labyrinthine that it has been claimed that no one understands fully both the act and the various regulations promulgated to implement it."[86] In the area of air pollution, for example, the federal government has established both new source performance standards and air-quality standards. Special requirements govern the pollution control technology required of both new plants and the expansion of existing facilities in both "prevention of significant deterioration" (PSD) and nonattainment areas; the latter are particularly complex, requiring both elaborate models to predict the impact of investments on air quality and a system of "offsets." State implementation plans may contain additional requirements.

An EPA report citing the 175 permits that Standard Oil of Ohio was required to receive before constructing a pipeline between Southern California and Texas—a project ultimately abandoned—noted:

> The complex of environment laws that exists today was formed incrementally over time; each new law was passed to address a specific single purpose or need, and subsequent laws were passed to fill in gaps left uncovered by the old. Moreover, organizationally separate agencies and programs also developed incrementally at the local, state and federal levels. As a result of this history, these agencies frequently have overlapping, duplicative or contradictory regulatory authority, as well as inadequate communicative networks.[87]

Another study reports:

> The most difficult and costly permitting delays for major new sources occur primarily in the preconstruction approval process for PSD areas. ... Add to this the requirement to furnish one year's air quality monitoring

data with the application, and the time involved between the submittal of an application and the receipt of a permit usually ranges from 10 to 21 months and the typical delay for the entire permitting process ranges from two to three years. The direct cost of paperwork, monitoring and analysis of air quality has been established at $250,000 to $500,000.[88]

A Brookings study adds: ". . . industrial critics of the program point to the paperwork, the extensive modeling required to both establish air quality baselines and to project incremental emissions, and the inappropriate and arbitrary nature of the emissions requirements themselves." More generally, the study notes that "no aspect of the Clean Air Act is more frustrating to industry than the expensive analysis, multiple delays and uncertainties in the permit process for construction or modification of plants."[89] Moreover, even if a firm has been granted a permit either to construct or to modify a facility, the company has no guarantee that emissions requirements will not be changed in the course of the plant's life.

The complexity of regulations governing the construction of new facilities in the United States does not simply create additional paperwork for companies and lead to considerable delays and uncertainties; it also undermines the legitimacy of siting decisions once they have been made. Major new projects have been denied approval in various parts of the United States, including a Dow Chemical facility planned for northern California, a Standard Oil of Ohio pipeline planned for the Southeast, and an oil refinery and marine terminal planned for Portsmouth, Virginia, near the mouth of the Chesapeake Bay.[90] What is significant about these controversies is not that planning approval was denied; as we have seen, such denials have also occurred in Great Britain. It is rather that these denials were attributed not to the greater importance officials had decided to place on amenity values but to the complexity and arbitrary nature of the procedures themselves. Siting decisions are more likely to be accepted as legitimate by industry in Britain—even when they go against a particular developer—because they are perceived to have been decided on their merits. In America, on the other hand, substantive issues often tend to be overshadowed by procedural ones.

In addition, because each planning application in Britain is considered on an ad hoc basis, the outcome in any particular case does not necessarily establish a precedent for future public policy. In the United States, however, because the approval of new facilities is in principle governed by the application of a complex set of detailed rules, each denial of a permit is seen as establishing a precedent for future governmental restrictions, thus adding to the tension between regulatory authorities and business. This certainly was the case in the permit denials cited above, two of which became national causes célèbres.

The British government has taken considerable pains to avoid the duplication of regulatory functions among governmental units. In the area of air pollution, the division of authority between the Alkali Inspectorate and local public health authorities is clearly defined: "registered" firms and "scheduled" processes are regulated by the former, all others by the latter. In the case of water pollution the situation is even simpler: all emissions are under the jurisdiction of one of ten regional water authorities. While the jurisdictions of governmental units responsible for pollution control and those of local planning authorities overlap, the Department of the Environment has made every effort to keep their respective regulatory responsibilities as distinct as possible.

It is true that on occasion the processing of planning applications has been considerably delayed. During the first half of the 1970s the DOE faced a large backlog of appeals from the decisions of local planning authorities. It was three years before approval was finally granted for the construction of an LNG terminal off the coast of Scotland, and the delays in choosing a site for an additional international airport facility for the London area have been extensive, to say the least. But on the whole, land-use decisions tend to be made relatively rapidly in Britain. In most cases the process of decision making is relatively straightforward, involving usually only two "rounds," one at the local level and another at the national level. (For those applications that have been "called in" there is only one round.) Moreover, the amount of documentation that firms are required to submit to pollution-control authorities before constructing new facilities is only a fraction of that needed in the United States. Great Britain does not require the preparation of environmental impact assessments, and with the exception of the large public inquiries, which rarely involve private companies, the hearing process itself is relatively informal and nonadversarial. British pollution-control requirements are not subject to judicial review, as they are in the United States, are rarely affected by legislative action, and do not require public hearings before they can be promulgated; in addition, inspectors enjoy considerable discretion in negotiating the terms of consents and the establishment of presumptive limits with particular firms. As a result, the regulatory process moves rather expeditiously.

*Implications*

This analysis helps explain why the American policy of strict enforcement has been no more effective in changing corporate behavior than

Britain's emphasis on voluntary compliance.[91] The more adversarial mode of regulation adopted by the United States makes voluntary compliance by industry problematic; to the extent that regulations are regarded by the private sector as unreasonable, firms are likely to comply with them only if they are forced to do so. For this reason "considerable effort and resources must be invested in controlling and monitoring activities, as well as in the prosecution and citation of violators, if regulations are to be effectively enforced." Such an effort is not only costly; its effectiveness is also limited. In general, "the practical limits to any further extension of inspections and more strict citations is often reached before the number of violations has been reduced to a minimum and the risk associated with noncompliance has therefore been minimized."[92] Moreover, the greater reliance of American officials on prosecution has not only made enforcement both time-consuming and expensive but also made it more problematic: not all the suits filed by the EPA have been settled or decided in the agency's favor.

On the basis of a comprehensive study comparing controls over sulfur emissions in ten European countries, Peter Knoepfel and Helmut Weidner conclude:

> Evidence from the United Kingdom suggests that compared to open systems (such as the United States), countries with closed enforcement processes are more likely to achieve a higher degree of compliance with the license conditions formally stipulated. The close cooperative relationships between the agency and the emitter, which are not disturbed by public participation, may, paradoxically, even lead to environmentally more favorable results in those cases where energy savings for the firm can be realized at the same time. In contrast to more open systems, such decisions, which are often combined with the introduction of complex technologies, can be carried out without great administrative costs.[93]

Certainly there are few firms in Britain whose pollution-control efforts have been as niggardly as those of the western smelters or midwestern utilities. American regulations generally demand more than their British equivalents, but more often than not the result is to make implementation in America more contentious, not to secure a greater improvement in corporate behavior.

The expanded opportunities for public participation in the making and enforcement of environmental regulations in the United States have proved to be a two-edged sword. For at the same time that they have granted environmentalists the right to challenge agency decisions, the courts also have expanded the opportunities available for industry to delay or resist compliance.[94] While environmentalists have been able to use litigation to require the EPA to make its regulations stricter, industry has also used the increased willingness of the courts to review administrative decisions as a way of challenging their legality in the first place. As a result, the EPA has found it far more difficult and time-consuming to establish pollution-control requirements for particular industries and processes than its British counterparts. The increase in congressional involvement in the regulatory process has led to a similar result: it has forced the EPA to adopt stricter rules and deadlines, but it has also provided industry with an additional vehicle for delaying their implementation; firms can petition Congress to change or amend the original legislation.

More generally the American experience demonstrates that overregulation can readily lead to underregulation. By requiring the EPA to accomplish so much, Congress has virtually ensured that the agency's rule-making and enforcement efforts will be inadequate: the price the EPA has paid for its successful effort at strict enforcement in selected policy areas—such as the control of automobile emissions and new source performance requirements—has been the virtual neglect of its regulatory responsibilities in a whole host of other areas. The latter include the regulation of hazardous air pollutants and potentially dangerous chemicals, the control of groundwater contamination, and the cleanup of toxic wastes. (For example, while the Toxic Substance Control Act authorized the EPA to regulate the use of 55,000 potentially hazardous chemicals, by 1980 it had managed to establish rules for only three.) In addition, the effect of statutes that force the EPA to establish strict rules—for example, Section 112 of the Clean Air Act of 1970, which requires the agency to establish standards for hazardous pollutants that protect the public health "with an ample margin of safety, and without regard for costs"—has been to reduce the number of hazardous pollutants that have been controlled at all: because "literal compliance with this criterion might require zero exposure in the case of carcinogens," by 1981 the EPA had succeeded in establishing standards for only four of the hundreds of potentially hazardous air pollutants.[95]

Broadly speaking, regulatory officials can choose one of two strategies to influence business behavior: they can pursue a policy of strict enforcement or one based on voluntary compliance. American environmental policy has tried to rely on the former, British environmental policy on the latter. The evidence presented in this chapter suggests that both have led to roughly similar environmental outcomes.[96] They have differed significantly, however, in their political implications; the former has produced far more conflict between business and government than the latter.

# CHAPTER FIVE

# Government Regulation in Great Britain and the United States

We are now in a position to appraise the significance of the similarities and the differences to be found in the British and American approaches to environmental regulation. This chapter examines the extent to which environmental regulation constitutes a unique case of government regulation in either society. Can we, in fact, use environmental regulation as a vehicle for generalizing about the legitimacy and effectiveness of government regulation of industry in Great Britain and the United States? Or are the differences we have described rooted in the particular characteristics of this issue?

Environmental regulation certainly exhibits anomalies in both nations. The Alkali Inspectorate, for example, does represent a vestige of Victorian public administration. Not only did it keep the same name for more than a century, but its current statutory authority derives from legislation approved by Parliament before World War I. Few dimensions of British regulatory policy are governed by such carefully defined rules of procedure or provide such extensive opportunity for public participation as does the British system of land-use planning, and few aspects of British regulation of industry have been so contentious over the last two decades.

Environmental regulation also exhibits anomalies within the context of American regulatory policy. No aspect of the wave of "new social" regulation enacted over the last two decades has produced so much political conflict. For more than fifteen years, conflicts over environmental policy have occupied a prominent place on the American political agenda. While the focus of public debate has changed since the late 1960s, the salience of the issue itself has not: the controversy over automobile emissions has given way to concern over acid rain, while the urgency of cleaning up the nation's lakes and rivers has been transformed into a concern over the disposal of toxic wastes. In no other area of social regulation have the attempts at "regulatory reform" initiated by the Reagan administration created so much conflict and controversy. And in no other policy area have the changes in the balance of power between industry and nonindustry constituencies been more substantial or proved more durable; the environmental movement entered the 1980s with more political strength than that of any of the public interest groups whose emergence so transformed business–government relations during the late 1960s and early 1970s. Compliance with environmental regulation has been a greater burden on industry than compliance with regulations in any other area; not surprisingly, in no other area of social regulation has the American business community devoted so many resources in an attempt to influence policy outcomes.

These distinctions persist if one turns to the actual dynamics of regulatory policy making. While Congress has come to play a much more important role in the policy process in every area of regulation over the last fifteen years, in no area has it become more involved than that of pollution control. The eight-volume legislative history of the 1977 Clean Air Act Amendments spells out in considerable detail the intent of Congress concerning a wide variety of matters that formerly would have been left to the discretion of agency officials. Similarly, while the federal judiciary has become much more involved in the making of regulatory policy, in no area of government–business relations other than antitrust activities has its role been so important as in environmental regulation. In addition, the EPA's rule-making procedures are more elaborate and legalistic than those of most other regulatory agencies.

The differences between British and American environmental policies may also be due to the unique characteristics of this issue. Thus the greater degree of political conflict over environmental regulation in the United States may simply reflect the greater saliency of amenity issues among the American citizenry. Whether owing to the heritage of citizen activism generated by the civil rights and antiwar movements or the actual magnitude of the nation's environmental problems or a combination of the two, the public certainly became far more preoccupied with government policy in this area in the United States than in Britain. The British public, lacking the experience of 1960s citizen activism and benefiting from the regulations adopted during the 1940s and 1950s, appears to have been less concerned with this particular dimension of business–government relations over the last two decades. The relatively strong performance of the American economy during the second half of the 1960s may have contributed to the raising of public expectations in this policy area, while Britain's persistent balance-of-payment difficulties and slower growth rates may have tempered public expectations and thus made the task of government officials easier. Finally, the greater degree of cooperation between industry and regulatory officials in Britain may in part be due to the fact that a relatively large proportion of the major industrial sources of pollution in Britain are owned by the government. In America, with the notable exception of the Tennessee Valley Authority, they are in the private sector.

In order to evaluate the significance of these factors, this chapter compares British and American regulatory policies and outcomes in three other policy areas: occupational health and safety, consumer protection, and the regulation of financial markets. Its conclusion is that environmental regulation does *not* constitute a unique case of government regulation in either society: the two nations regulate the impact of business decisions on the environment in much the same manner— and with substantially the same results—as they regulate a wide variety of other dimensions of corporate conduct. Moreover, neither the degree of public concern, the nature of ownership, nor the profitability of particular industries appears to be decisive in explaining either nation's regulatory style. It is rooted rather in the broader dynamics of business— government relations—a subject we shall examine in Chapter 6.

Students of comparative public policy disagree over the extent to which public policy processes and outcomes are due to factors unique to a particular policy area—the so-called policy sector hypothesis—or whether they are a function of the political and social characteristics of the nation in which they are developed—the so-called national differences hypothesis. Gary Freeman writes:

> The question can be stated very simply. Which is more important in the study of comparative public policy—the structure of politics in a particular country and the differences in such structures between countries (that is, the nature of the national *apparatus* for making public policy), or the characteristics of the political issues (the policy arena or sector) with which public policy is attempting to deal? Will countries with dissimilar political systems produce significantly different responses to similar political issues, or will the imperatives of particular types of public problems compel more or less similar responses, whatever the shape of the political system involved?[1]

The evidence presented in this chapter supports the conclusion that the characteristics of a political regime are more important than the nature of the particular policy area itself in explaining policy processes. On the other hand, both the policy agenda and policy outcomes appear to be more affected by socioeconomic variables. Accordingly, among advanced industrial societies, the former vary much more than the latter.

## OCCUPATIONAL HEALTH AND SAFETY

The area of regulation that most closely resembles environmental protection is occupational health and safety. As in the case of pollution control, the efforts of the British government to protect the health and safety of workers have a long history. The first factory act was enacted in 1833, and legislation regulating the working conditions of miners was enacted in 1850. Throughout the nineteenth century, a series of amendments to the Factory Regulation Act of 1833 steadily expanded its scope. The original legislation applied only to children, but by the turn of the century, factory legislation had begun to apply to many adults as well. Many factory inspectors, like local public health officials, were not technically trained. The technical qualifications of the Mines Inspectorate, however, did closely resemble those of the Alkali Inspectorate: it consisted—and still consists—of qualified mining engineers with considerable experience in industry.

From the outset the inspectorates responsible for safety at the workplace operated on the assumption that while inspectors might do all they could to see that manufacturers and mine operators complied with the law's technical requirements, the ultimate responsibility for improving the working conditions of employees rested upon the owners and managers themselves. Accordingly, the success of the various inspectorates was to be judged not by their ability to secure compliance with minimum legal requirements but by their success in persuading employers and managers to assume greater responsibility for the physical welfare of their employees. The critical role of industry itself in improving health and safety conditions for employees was repeatedly emphasized in the reports of the Factory Inspectorate. In his 1963 annual report, for example, the chief inspector noted that most industrial accidents were not connected with violations of factory legislation. Rather, they "occur in circumstances which cannot readily be controlled by legislation, for example, lack of attention to good industrial housekeeping and to general tidiness . . . and above all lack of knowledge and application of safe methods of work."[2]

While prosecutions for breaches of factory legislation were somewhat more common than those for violations of pollution-control regulations, they were still relatively infrequent: one study of an inspector's files between 1961 and 1965 revealed that although inspectors had uncovered

3,800 contraventions of the factory acts and their regulations, in nearly 75 percent of the cases "the inspector took action by means of a formal notification of matters requiring attention while in less than 4 percent did he either threaten prosecution directly or actually institute proceedings."[3] Another study reported:

> It is extremely rare for a general inspection *not* to result in a "discovery" and formal notification of a number of breaches of the law for which criminal proceedings could be instituted. But in the vast majority of cases such matters are not regarded as criminal offenses in any ordinary sense either by the Inspectorate or by the employees concerned.[4]

In the case of the Mines Inspectorate, a royal commission noted in 1938 that prosecutions, which had averaged thirty-two a year at the beginning of World War I, had fallen to nine a year by the 1930s. Between 1948 and 1975 they ceased altogether.

As in the United States, criticism of the effectiveness of occupational health and safety regulation increased substantially in Britain during the 1960s. While much of it came from individual workers, in both nations a significant source of pressure derived from trade unions. Unions in Britain wanted not only to increase the representation of workers in the administration of safety procedures at the plant level but also to strengthen the enforcement efforts of the Inspectorate. Because the Inspectorate continued to insist that the primary responsibility for providing a safe workplace rested with management, the unions suspected that too little attention was being paid to the views and needs of employees. Moreover, the activities of the Factory Inspectorate came under increased scrutiny as part of the public's heightened interest in environmental regulation, since the public's interest in its safety and that of a firm's employees in their own were identical.

In Britain, as in the United States, public concern was also heightened by a number of highly publicized instances in which the government's regulatory body appeared to have been negligent in protecting the health and safety of British workers. Her Majesty's Factory Inspectorate was criticized, for example, for failure to protect the workers exposed to asbestos at the Bermondsey Factory of Central Asbestos and to control the lead pollution produced by the RTZ smelter at Avonmouth. Both works were eventually closed—but not before many of their employees had been injured.[5]

Immediately before the general election of 1970, the secretary of state for employment and productivity appointed a committee under the chairmanship of Lord Robens to reexamine the entire question of safety and health in the workplace. Its mandate was extremely broad: it was not only to examine the entire question of safety and health at the workplace but to consider what was needed to protect the public from hazards arising from industrial production. The Robens Committee released its report in 1972. Its main conclusion was that the existing system of regulation relied too much on specific rules and regulations. It urged a more flexible statutory system that would encourage greater "personal responsibility and voluntary self-generating effort."[6] The committee also recommended that a new authority be established with exclusive responsibility for health and safety at work and that the six existing regulatory bodies concerned with this function be merged into a single unified inspectorate.

While explicitly rejecting the argument that what was most needed was a strengthening of the police powers of the Factory Inspectorate and greater willingness to prosecute employers, the Robens report did propose strengthening the administrative sanctions available to inspectors. Its primary recommendation, however, was that strict legal enforcement be deemphasized and that the role of inspectors as advisers to both employees and industry be strengthened. This position more closely reflected the preferences of the Confederation of British Industry than it did that of the Trades Union Congress. The latter recognized the importance of self-regulation but did not regard it as an alternative to stricter enforcement of safety legislation.

After considerable interdepartmental conflict—much of which centered on the extent to which each of the previously independent inspectorates should become part of one organization—the government proposed the Health and Safety at Work Act of 1974. (Some of the strongest objections came from the Royal Commission on Environmental Pollution, which unsuccessfully opposed the removal of the Alkali Inspectorate from the Department of the Environment to this new regulatory body.) A governmental spokesman for the legislation did concede that "there was anxiety and apprehension about the general philosophy of the report," which seemed to call for a retreat from strictly enforced statutory requirements.[7] However, the legislation submitted to and approved by Parliament accepted the report's main recommendations.

The 1974 legislation established two new bodies: a Health and Safety Commission (HSC), to be composed of nine members appointed by the secretary of state for employment after consultation with representatives of employees, employers, local government, and various professional associations, and a Health and Safety Executive (HSE), which was to be responsible for administering the policies made by the commission. The latter's responsibilities included information and advisory services, research, and inspection. The HSE was also responsible for establishing special working bodies, comprised of members drawn from the various bodies represented on the HSE, which would develop guides and codes of practice for particular occupational hazards. The new law also extended protection to the self-employed and provided for the protection of the general public from hazards to their health and safety from industrial activities. The latter provision involved factory inspectors for the first time in environmental regulation.

The contrast between the Health and Safety at Work Act (HSWA) and the Occupational Safety and Health Act (OSHA), signed into law by President Nixon on December 19, 1970, is marked. While both laws were enacted in response to considerable public and union dissatisfaction with existing regulatory arrangements, the American legislation was enacted after an extremely bitter legislative struggle. The unions originally supported a bill that would have given the secretary of labor the power to set and enforce standards as well as the right to shut down plants where an "imminent danger" was found to exist. After considerable opposition from Republican legislators and business lobbyists, a compromise was reached. The final version of the bill placed the power to set standards and review violations in a separate body, the Occupational Safety and Health Administration (also known as OSHA), which was placed under the jurisdiction of the Department of Labor. More important, while the British statute essentially reinforced existing trends in British regulatory practice—its primary innovation was organizational—its American counterpart represented a radical policy departure in two significant respects. Like the National Environmental Policy Act of 1969, the Clean Air Act Amendments of 1970, and the Clean Water Act Amendments of 1972, the Occupational Safety and Health Act transferred the primary responsibility for the making and enforcement of regulations from the states to the federal government: states could continue to administer their own regulations, but only if they were at least as stringent as those promulgated by OSHA. Second, it reflected a shift to a more adversarial and rule-oriented approach to policy making and enforcement.

The two nations' regulatory strategies thus differ substantially. The British have moved in the direction of increasing the responsibility of private parties for designing and maintaining safe conditions at work. Thus the HSWA contains a series of "general duty" clauses that define the respective obligations of employer, self-employed persons, managers, and employees to ensure safety and reduce health risks in the workplace. Consistent with the recommendations of the Robens Committee, over the last decade the Factory Inspectorate has found itself increasingly drawn into "develop[ing] new techniques for monitoring what the employer does, appraising safety organizations rather than simply seeing that employees conform to specific rules."[8] In addition, the 1974 legislation formally institutionalized a tripartite system that enables worker representatives as well as employers to participate in the development of standards and codes of conduct. In fact, the HSC routinely ratifies the standards approved by labor and management negotiators—each of whom has the power of veto.

The American statute, by contrast, placed the authority to promulgate occupational safety and health standards exclusively in the hands of the Department of Labor—though subject to the formal procedures of administrative rule making. By turning over the task of establishing standards to the representatives of the interest groups directly affected by them, the British system reduced the direct role of government in the regulatory process; by establishing a new agency responsible for promulgating workplace health and safety standards, the American approach increased it. While the HSWA "reflects well-known features of British regulatory policy in favoring a decentralized, flexible development of rules, which apply ordinarily to specific employers rather than entire industries," the OSHA "calls for rules to be promulgated by a central government agency to be uniformly applicable to whole classes of employers."[9] And while OSHA has severely restricted the autonomy of its field inspectors, encouraging them to "go by the book," the British system continues to provide inspectors with substantial discretion.

The Occupational Safety and Health Act encouraged penalties both by allowing fines to be imposed through a "civil penalties" procedure that required a relatively low standard of proof and by requiring inspectors to cite employers for every breach of a regulation that they observed (whether or not it actually threatened workers' health or safety). British regulators, now as in the past, rely more on conciliation than on coercion to achieve policy objectives and employ legal action only as a last resort when persuasion proves inadequate. In 1980, 140,000 fines were imposed on coal-mining companies for health and safety violations in the United States, while in Britain there have been no prosecutions for more than two decades. Although the number of prosecutions by the Factory Inspectorate has increased since 1974, British inspectors continue to "prefer advice and persuasion to the big stick of the criminal law." As the 1979–80 Annual Report of the HSE put it: "In raising the level of workplace health and safety awareness, and encouraging a more positive approach to combating hazards, we have proceeded . . . on a basis of agreement."[10] Charles D. Drake and Frank B. Wright add:

This consensual approach operates at the national level using consultation on regulations and codes of practice and relying on the advice of advisory committees and at the plant level by consulting safety representatives and safety committees. The new summary powers to issue improvement and prohibition notices backed up by the ultimate discretion to prosecute are seen as reserve powers to coerce the recalcitrant.[11]

By contrast, Steven Kelman writes:

Perhaps the best single word to describe the American enforcement process, in its various aspects, is *formal*. OSHA lays out inspection procedures in great detail. Inspections take the form of searches for violations. They result in formal citations (and penalties) for specific regulatory infractions. These citations are frequently formally contested by employers. If contested, OSHA must formally prove, in a court setting, that the violations did in fact occur. The process of enforcing the rules thus itself goes according to well-defined rules.[12]

The two nations also differ in their approaches to the weighing of costs and benefits in the setting of standards. The British legislation mandates that workers should be protected "so far as is reasonably practicable"—a phrase that represents the rough equivalent to the criterion of "best practicable means" employed by the Alkali Inspectorate in establishing emission requirements. As in the area of pollution control, inspectors, acting within the guidelines established by the working parties established by the HSE, are responsible for balancing what is appropriate on safety grounds against the degree of risk and the cost of prevention. The British system of regulatory decision making obviates the need for a formal system of cost-benefit analysis. Sheila Jasanoff notes:

Accustomed to negotiating with each other, neither labor nor industry requires the other's demands to be spelled out in monetary terms, and the possibility of bargaining on policy outcomes removes the need for explicit principles of decision. Because government ratifies agreements among the relevant private interests, rather than making choices of its own, there is little reason for regulators to justify policy decisions through economic or other technical arguments. The consultative documents on major regulatory policies often include some discussion of costs and benefits. But the analysis is more qualitative than quantitative, neither pretending to excessive methodological rigor nor serving to justify a particular policy choice.[13]

In the United States, on the other hand, the use of cost-benefit analysis in the making of regulatory decisions has been the subject of considerable controversy. The statute itself is ambiguous: one section appears to encourage the weighing of costs and benefits in formulating specific regulations while another appears to prohibit it. In the one Supreme Court decision that has addressed this issue, the "cotton-dust" case (*American Textile Manufacturers Institute* v. *Donovan*) the Court held that the phrase "the extent feasible" in section 6(b)(5) superseded the "reasonable" requirements of section 3(8). The majority argued that "feasible" meant "capable of being done," and that if Congress had wanted costs to be taken into consideration when standards were formulated, it would have said so in the statute.[14] Whether or not the Court's decision applies solely to those regulations promulgated under rule 6(b)(5) or to all OSHA regulations remains unclear. In the early 1980s, British officials did become more concerned about the cost-effectiveness of particular regulations, and the HSE hired its first economist. But the use of cost-benefit analysis remains far less contentious in Britain than in the United States. Whereas American regulatory officials have come under substantial pressure to establish standards that "guarantee" a safe workplace, the British public appears to accept a certain amount of occupational injury as a necessary concomitant of industrial production.

## VINYL CHLORIDE

The contrasts in the British and American approaches to protecting health and safety in the workplace emerge with particular clarity in the two nations' approaches to the regulation of vinyl chloride (VC), a chemical that is the cornerstone of the plastics industry in both the United States and Western Europe. In January 1974 B. F. Goodrich announced the deaths of three of its workers who had been exposed to vinyl chloride and formally informed the National Institute of Occupational Safety and Health (NIOSHA), OSHA's research arm, that VC appeared to be linked to cancer in humans. Within a relatively short time, Goodrich's finding had become available to government officials, corporate managers, and trade unionists in both Western Europe and Japan. The issue received considerable international media attention and both government and industry throughout the industrialized world found themselves subjected to substantial public pressure to reduce the exposure of workers to what was clearly an extremely hazardous substance.

The first country to act was the United States. On January 30, 1974, OSHA announced in the *Federal Register* a fact-finding hearing on VC to be held the next month. The purpose of the hearing was to determine if OSHA should issue an emergency temporary standard (ETS) immediately to protect workers from "grave danger" or if it should follow its regular rule-making procedures. Industrial spokesmen vigorously argued against an ETS, contending that they had already reduced average exposure to below 50 parts per million (ppm), a level at which no threat to human health had been demonstrated. Union officials, on the other hand, insisted that VC was clearly a human carcinogen and that the government should immediately require companies to eliminate all worker exposure. On April 5 OSHA issued an ETS that specified that no worker could be exposed to more than 50 ppm. A month later OSHA issued its "Proposed Standard for VC," which proposed that VC exposures be set at the "nondetectable level, as determined by a sampling and analytical method capable of detecting VC at concentration levels of one ppm." Companies had to comply "as soon as feasible."

This regulation fell, in the words of one industry official, like a "bombshell." One executive noted: "The standard made a lot of people in the industry furious. I mean mad enough to start fist fights. People said, 'They can't do that to us—we're running a clean operation.' When further OSHA announcements came out, the attitude was, 'Did you see what those bastards did to us today?' " The trade association that represents most major plastics producers in the United States convened a series of special meetings to plan its political and legal strategies. After considerable intra-industry conflict, the producers decided to recommend an average exposure level of 25 ppm, to be effective in October 1974, and a level of 10 ppm to take effect two years later. Industry officials publicly attacked OSHA's proposed standard, calling it "impractical," "unachievable," and "a disaster for industry." Union officials counterattacked: an official of the Oil, Chemical and Atomic Workers Union stated that "workers are tired of being used as guinea pigs." The president of the United Rubber Workers sent a message to President Nixon, urging that the government "stop the cruel and unnecessary killing of our workforce."[15]

In June and July, OSHA held a total of eight days of hearings. The official record exceeded 4,000 pages and included more than 800 oral and written submissions. While industry representatives contended that recent deaths were due to the very high levels of exposure that had prevailed in the industry ten or twenty years earlier, cancer specialists from NIOSHA and the National Cancer Institute testified that no threshold level had been established for human carcinogens. Industry officials claimed that OSHA's standards were not economically feasible, while various union officials and representatives of public interest organizations disputed this claim. In October 1974 OSHA published its final standard: it set "an exposure limit of 1 ppm average over any eight-hour period and a ceiling of 5 ppm average over any period not exceeding five minutes."[16] While labor's response was lukewarm, within two minutes of OSHA's announcement seven firms had filed suits in various federal courts challenging OSHA's ruling. These appeals were unsuccessful and the industry then set about seeking to comply. Within two years most firms were in compliance; the total costs to industry were estimated at between $200 and $280 million. As this cost was substantially less than had been predicted, critics accused the firms of having cried wolf.

The chief inspector of factories in Great Britain was informed of B. F. Goodrich's announcement almost immediately after it was made. Within a week, Imperial Chemical Industries (ICI), the major British chemical firm, had furnished the government, officials of the Trades Union Council (TUC), and its own employees all the information on VC it had available. Following a series of meetings among industry scientists, the TUC's medical adviser, and the senior government officials responsible for occupational health, the two major British chemical producers, British Petroleum Chemicals and ICI, voluntarily agreed on an initial objective of reducing VC levels to below 50 ppm. A few months later the Chemical Industry Association established a special VC Committee to coordinate the technical research efforts of the industry and to work with the VC working group of the HSC. (In fact, the membership of these groups overlapped.)

In the spring of 1974, the Factory Inspectorate, seeking to comply in advance with the provisions of the Health and Safety at Work Act—which was then in its final stages of parliamentary approval—invited industry and labor to form a joint working group to develop an industry code of practice. The group, chaired by the chief inspector of factories, consisted of five labor representatives and an equal number from industry; its staff included a medical statistician, the deputy chief of the Alkali Inspectorate, and representatives of the Department of Health and Social Services. All of its meetings were held in private, and while its members kept their constituencies informed of its work, contacts with the press were extremely limited. In December the committee approved a draft code of practice, which established an "interim hygiene standard," defined as

> a ceiling value of 50 parts per million, that must not be exceeded; the average exposure over the whole shift must not exceed 25 parts per million. These are outside limits and it is further asked that exposure should be brought as near as practicable to zero. This standard, which was to be kept under review by the Working Group, therefore encourages progressive reductions of exposure to the lowest possible levels given present engineering knowledge.[17]

Members of the committee subsequently made a series of visits to VC plants in order to discuss the hazard informally with workers and examine the monitoring and control technology available to industry. These visits served the additional purpose of developing close personal relations among the members of the group. An official of the Transport and General Workers Union noted that while initially both industrialists and labor representatives viewed each other with suspicion, after observing industry efforts to control VC, the participants achieved "a real breakthrough in relations" and got along "famously"; industry members in turn described their trade union counterparts as "level-headed and responsible men."[18]

In May 1976 the group approved a final code of practice. It established a "ceiling value of 30 ppm in a time weighted average of 10 ppm," with the understanding that "whenever practicable exposure should be brought as near as possible to zero concentrations."[19] This code had no legal status: it was simply a voluntary agreement. Even before it was announced, however, average weekly VC levels in British plants had been reduced to somewhere below 5 ppm, and by 1977, plants in Great Britain were maintaining levels between 0.5 and 3.0 ppm. The capital costs of compliance for British industry as a whole were estimated at more than £20 million.

Thus, while the exposure level of workers in the United States may have declined somewhat more rapidly, within three years VC exposure levels were virtually identical in the United States and Great Britain. The ways in which this objective was achieved, however, differed dramatically. The United States established a strict, technology-forcing standard that left no room for variation among producing firms in controlling VC exposure levels; all had to comply with uniform requirements as soon as possible. The British, on the other hand, deliberately set the standard *above* the lowest level possible with available technology, and companies were able to calculate costs and benefits individually. They could decide on a plant-by-plant basis whether further reductions in exposure made good economic sense. As a result, while the effective exposure levels tended to be relatively similar among all plants in the United States, exposure levels in Britain tended to be higher in some plants than in others.

## EVALUATING HEALTH AND SAFETY REGULATION

As in the case of environmental regulation, Britain's efforts in the area of occupational health and safety have become more controversial over the last decade. The inability of government and industry to provide adequate protection to workers exposed to asbestos, for example, has been persistently criticized. The first British regulations on occupational exposure to asbestos dust were issued in 1933. In 1968, responding to research by the British Occupational Hygiene Society, the maximum occupational exposure was reduced to 2 million fibers per cubic meter. This standard was attacked as "grossly inadequate," however, and in 1982 the Health and Safety Commission acted to reduce it by half. The asbestos industry, which was already operating at the 1-fiber standard level, vigorously opposed the government's efforts to reduce it to 0.5 fiber. It spent £500,000 on a public relations campaign designed to "influence reports, legislation and government regulations on new safety standards."[20] A report issued by the British Society for Social Responsibility in Science countered by predicting that 500,000 workers might die from asbestos-related diseases over the next thirty years.

In the late 1970s the National Union of Agricultural and Allied Workers challenged the government's unwillingness to ban the use of the pesticide 2,4,5-T, "claiming that a terrifying trend of abortions, miscarriages, and infant deformities in the families of agricultural workers can be related to [its] use."[21] The government, however, agreed only to restrict its use. A resolution passed at the 1980 TUC Annual Conference termed the voluntary registration scheme a failure and urged that the control of pesticides be removed from the Advisory Committee on Pesticides and placed in the hands of the HSC.

In 1982 W. G. Carson published *The Other Price of Britain's Oil*, which sharply criticized the safety standards being applied to workers on oil-drilling equipment in the North Sea. While acknowledging that the accident rate from the British sector of the North Sea was lower than that for the Norwegian sector, Carson contended that a significant number of accidents to British workers could easily have been avoided had safety considerations been accorded a higher priority by both the Department of Energy and the oil industry.[22]

In comparison with the turmoil generated by health and safety legislation in the United States, however, the controversy in Britain has been relatively modest. In the United States, dissatisfaction with occupational health and safety regulation has been pervasive in the ranks of both labor and industry; indeed, no other regulatory agency has been so universally unpopular as OSHA. Industry has accused it of being obsessed with enforcing detailed rules that often bear no relationship to actual hazards, overeager to prosecute, indifferent to the costs of compliance, and contemptuous of industry expense and expertise; indeed, critics of OSHA have argued that in spite of its zealousness, it has actually done nothing to improve occupational safety. Trade unions, on the other hand, have accused OSHA of being far too lax in both setting and enforcing standards.

As in the case of pollution controls, measuring the impact of regulation on health and safety at the workplace is extremely difficult. Workers' health and safety are affected much more by their ages, their turnover rates, and the type of production than by government regulation. On balance, the British appear to have been slightly more effective in reducing employee injuries. Between 1974 and 1980 the incidence rate of fatal injuries in Great Britain fell between 4.5 per 100,000 workers at risk to 2.7 (from 16.0 to 13.0 in construction). The incidence rate of reportable accidents per 100,000 workers at risk fell from 3,570 in 1976 to 2,860 in 1980 (from 3,410 to 3,000 in construction). On the other hand, the consensus of those who have analyzed the data is that during OSHA's first decade the agency had a negligible effect on the occupational injury rate in the United States, though since 1980 the rate has fallen.[23] While the Mine Safety Program did bring about a significant decline in coal mine fatalities in the United States, the American fatality rate per mine-worker-hour is currently three times higher than the British.

Regulatory officials in both societies have placed less emphasis on the more difficult challenge of protecting the health of workers, but the efforts they have made have been roughly comparable. While the two nations currently have comparable standards for workplace exposure, British authorities identified asbestos as a hazard thirty years before the danger was recognized in the United States: it was not until 1964 that the American asbestos industry agreed to place labels on its products. The British also adopted a standard of 2 million fibers per cubic meter for workplace exposure four years earlier than did the United States. On the other hand, American officials have placed greater restrictions on the use of pesticides, thus providing more protection to agricultural workers. Because the resources available for scientific research on new workplace hazards are much greater in the United States, American authorities now appear to be in a better position to identify new sources of occupational disease at the workplace, although, as we have seen in regard to vinyl chloride, the results of such research are readily made available to regulatory authorities in Britain. Since, however, fatalities due to exposure to sources of occupational illness may not appear for many years, it may be some time before we can fully access the relative effectiveness of British and American regulatory efforts in this area. (As things now stand, the much more important role played by private lawsuits in the area of occupational illness in the United States may, in the long run, have a greater deterrent effect on corporate conduct than either nation's formal system of regulation.)

As in the case of environmental protection, one reason that the stricter American occupational safety and health standards have not brought about a more rapid reduction in accident and disease rates has been the difficulties the government has faced in implementing them. Largely because of industry opposition, OSHA was able to issue only four major health standards during its first six years. While the number of both health and safety standards issued increased somewhat between 1977 and 1980, so did industry resistance to their enforcement: in 1980, approximately 25 percent of all violations were contested by industry. Graham Wilson argues that OSHA might be able to issue regulations more speedily if it adopted an approach less likely to provoke opposition from industry. He asks:

> Might not a slightly less than perfect regulation which industry might accept be preferable to a perfect regulation which, like OSHA's benzene regulation, was invalidated after five years work, leaving the agency with nothing? As is often the case, a decent argument can be made for accepting compromise rather than defeat.[24]

These comparisons of British and American practices and policies in regard to the protection of workers from disease and injury are particularly significant because the political and organizational contexts of environmental protection and occupational health and safety policy differ so substantially in the two countries. The American public, on the whole, appears to have been more preoccupied with environmental regulation than the British. In addition, in comparison with their counterparts in the United States, British environmental organizations have fewer legal and organizational resources with which to participate in the policy process, particularly in areas that are highly technical.

These contrasts are much less striking in the area of occupational health and safety. Unlike British environmental organizations, the British trade union movement is relatively well staffed and funded. Not only is a far larger proportion of industrial workers unionized in Britain than in the United States, but those workers enjoy considerably more access to the regulatory process. While American trade unions function primarily as litigators and lobbyists, British union officials fully participate in the actual making of the occupational health and safety regulations that fall under the jurisdiction of the HSE. Yet despite these differences, both the contrasts and the similarities we have observed in the environmental regulations of the two countries can also be found in the area of occupational health and safety regulation. And as both the vinyl chloride case and our comparative accident data suggest, the more cooperative British approach to compliance has proved at least as effective as the more adversarial enforcement strategy adopted by officials in the United States, if not more so. It has, however, produced far less political conflict between industry and government.

## Chemical Regulation

These differences in occupational health and safety regulation in Great Britain and the United States extend to the more general area of health and safety regulation. An exhaustive study of chemical regulation and cancer in the United States, Great Britain, France, and the Federal Republic of Germany concludes that in virtually every case, U.S. standards tend to be relatively specific and detailed, while British rules tend to be much more flexible and informal. Thus while the United States makes extensive use of specific tolerance standards, "with equal regularity Britain avoids obligatory, legally enforceable controls of any type, except in the case of food additives and a few other specialized instances" (some of which were designed to conform with EEC directives).[25] And when the British do prescribe exposure limits, they generally take the form of recommendations rather than requirements.

> In the view of British public authorities, their non-statutory approach to the control of pesticides and workplace chemical hazards succeeds because of a high degree of compliance with voluntary standards on the part of industry. In the United States, in contrast, it is traditionally assumed that the behavior of those regulated requires close scrutiny under threat of punitive sanctions.[26]

Not surprisingly, British industry has generally been quite satisfied with its government's approach to the regulation of chemicals. By assessing and controlling chemical hazards substance by substance and by tailoring test requirements to specific chemicals, industry avoids the need for unnecessary and costly testing of new substances—frequently cited as a major bar to innovation in the United States. In contrast to the United States, which has devised highly specific regulations governing substances suspected of causing cancer—the most well known is the Delaney Amendment, which requires the Food and Drug Administration (FDA) to ban any food additive suspected of causing cancer—Britain has tended to subsume the regulation of carcinogens within its existing regulatory framework. And whereas American regulatory officials have on occasion banned particular substances or proposed drastic reductions in their use, the British prefer to cut back production gradually and to introduce restrictions on use. A comprehensive cross-national study of chemical regulation concludes:

> Managers of individual firms and officials at the Chemical Industries Association, the industry's trade association, enjoy close, even cozy relations with civil servants at the . . . Health and Safety Executive and the Department of the Environment. Informality is the way of doing business in Britain, founded on a spirit of trust between industry and government that permits the tackling of tough policy problems without public battles. . . . In sum, as one industry spokesman put it, if British chemicals must abide some form of government intervention for regulatory purposes, they could hardly design something superior to the existing approach.[27]

When the British system for the control of hazardous substances is understood in its broadest terms, the tripartite arrangement for controlling exposure to toxic substances in the workplace established under the auspices of the HSE is seen to be highly unusual. Two British scholars have surveyed the nature of the advisory bodies established to make decisions with respect to the regulation of dangerous chemicals in the workplace and in other policy areas: foods and food additives, medicines, pesticides, tobacco, consumer products. In every case but the first, the governmental body charged with formulating regulatory policy consists exclusively of "technically trained individuals who are selected not as representatives of any particular economic interest or political viewpoint but rather as independent experts who are able to evaluate the scientific evidence." Moreover, none of the bodies publishes transcripts of its deliberations or publicizes its agenda or minutes, and "each advisory body enjoys a high degree of discretion in selecting the form of recommendations which it deems suitable to specific circumstances."[28]

While some bodies, such as the Food Additives and Contaminants Committee, tend to follow relatively rigid guidelines, in general "standards are [not] rigidly enforced but are weighed against economic, industrial, or geographic conditions." Each advisory body tends to make a relatively sharp distinction between "insiders," who are able to participate fully in its deliberations, and "outsiders," members of the general public, whose access to the policy process is virtually nonexistent. Government officials justify this arrangement on the grounds that "it encourages . . . informal co-ordination and flexibility of approach in a manner which a more open and adversarial system (such as that found in the

United States) would not be able to achieve."[29] In this sense, the distinction between the decision-making procedures of these bodies and that of the Health and Safety Executive is roughly analogous to that between the Alkali Inspectorate and the British system of land-use planning.

Do the advantages that industry enjoys because of its close ties to the scientific community and the limited ability of nonindustry interests to challenge regulatory policies publicly expose British workers and consumers to unnecessary hazards? Certainly "the fortunes of the chemical industry are more favorably balanced against the need to protect health and the environment in Britain than elsewhere." Yet at the same time, when one turns from the way in which policy is made and implemented to specific policy outcomes, one finds that British and American regulatory policies are remarkably similar. Thus the use of all fourteen substances identified as carcinogenic by the highly respected International Agency for Research on Cancer has been restricted by regulatory authorities in both countries; in most cases these restrictions are identical. Despite the initiative shown by American regulatory officials and the stricter requirements of American law, the United States "has regulated few suspected carcinogens that do not also appear on the lists of controlled substances in Europe."[30] While the United States has moved somewhat more aggressively to restrict the use of food additives and some pesticides, the British have tended to place tighter controls on occupational carcinogens.

## DRUG REGULATION

The dimension of British health and safety regulation that has attracted the greatest interest in the United States is that of prescription drugs. The inspection of the quality of medicinal products in England dates from the reign of Henry VIII. In 1858 Parliament established quality-control standards for medicines and in 1875 legislation was enacted fixing penalties for the adulteration of drugs. The Therapeutic Substances Act (1925) provided for the licensing of medicines and established standards for the manufacture or import of substances "the purity or potency of which cannot be adequately tested by chemical means."[31] Following the thalidomide disaster in 1963, a Committee on Safety of Drugs (CSD) was established by the Ministry of Health after consultation with both the medical and pharmaceutical professions and the pharmaceutical industry. The committee established a voluntary regulation system similar to that adopted a few years earlier to regulate the production and use of pesticides. Under this system, approximately six hundred companies, most of which belonged to either the Proprietary Association of Great Britain or the Association of the British Pharmaceutical Industry, agreed to submit their products for scrutiny before conducting clinical trials or marketing; a system was also established to monitor approved drugs for adverse effects.

This scheme of self-regulation was subsequently changed. In 1968, faced with evidence that many British companies were not participating in the voluntary regulation scheme, Parliament enacted the Medicines Act, which established a mandatory licensing system to be administered by the Department of Health and Social Security. A Medicines Commission, composed primarily of professional experts, was formed to advise the minister on its implementation. Of several expert committees that were formed, the two most important were the Committee on Safety of Medicines, which advises the licensing authority on questions of the safety, quality, and efficiency of medicines for human use, and the Committee on the Review of Medicines, which reviews the safety, quality, and efficiency of existing products on the British market. Despite this shift from a voluntary to a statutory system of regulation, the pattern of industry–government cooperation established earlier in the decade has persisted.

> The Committee on Safety of Drugs and the voluntary system did set the pattern of informality of approach and good relationships with the industry that the Committee on Safety of Medicines and the Licensing Authority have striven to maintain. This "approachability" and flexibility of the British licensing system, which is much appreciated both by industry and by the medical profession and is admired throughout the world, is largely due to the attitudes engendered by the Committee on Safety of Drugs during its period of operation from 1964 to September 1971.[32]

Ironically, while the British system of drug regulation has been frequently cited by both academic and industry critics of American regulatory practices as a model for the United States, in fact the British legislation of 1968 has been subject to many of the criticisms leveled at the 1962 amendments to the Food and Drug Act. A survey of the British pharmaceutical industry found that the shift from a voluntary to a legally mandated licensing system

> had *not* resulted in a substantial improvement in patent safety. Firms were ... unanimous that the Act has had *harmful* effects through delays in the introduction of new drugs.... The industry claimed that the 1968 Act has had a major adverse effect on its R and D [research and development]

activity. The result is longer development, less innovation, less basic research, and more clinical R and D being undertaken abroad.[33]

These criticisms are similar to those made of the FDA's policies since the passage of the Food and Drug Act Amendments of 1962.

Nonetheless, substantial differences remain between the two nations' approaches to drug regulation—differences that parallel those we have observed in other areas of regulation.[34] In 1981 the average approval time for a new drug in Great Britain was thirteen months; in the United States it was twenty-three months. The United States figure actually represented a considerable improvement over 1979, when the average delay was thirty-five months; nonetheless, the average gap (dubbed the "drug lag") between the approval of a drug in Britain and its approval in the United States has remained constant at fourteen months. During the 1970s, nine of the twelve most important new drugs introduced in the United States were available an average of two years earlier to patients in Britain. According to one drug company, Eli Lilly, the seven best-selling drugs it introduced between 1970 and 1979 were approved for use on average twenty-three months earlier in Great Britain. The world's best-selling drug, Tagamet, an anti-ulcer medication developed by Smith-Kline Corporation, was approved by British authorities in two months; the FDA took thirteen months to process the application.

Moreover, the FDA's procedures for approving new drugs are far more burdensome than those of the Committee on Safety of Drugs. Thus while the FDA requires that its permission be obtained before even the first phase of clinical testing begins, British regulatory authorities require only that they be notified. The FDA has far stricter standards as to what constitutes an adequate and well-controlled trial than does the CSD. It also requires considerably more documentation: the application for the approval of Tagamet in the United States consisted of 160 volumes, whereas in Britain only 12 were required. On the other hand, the British have a far more extensive system for monitoring adverse reactions following approval than do the Americans, and have not hesitated to suspend drug licenses when ill effects have become evident.

There is no evidence that the British system of drug regulation provides less protection for the public than the far stricter American system: dangerous drugs, such as Oraflex, continue to be approved for use occasionally in both countries. On balance the American public appears to be worse off, since it also suffers from the FDA's delays in approving the use of beneficial new drugs. According to one study,

of the 82 drugs that became mutually available, 14 were introduced in the same year in both countries, 43 were introduced in Britain first, and 25 were introduced in the United States first. Britain enjoyed 120 drug years of prior availability, while the United States only enjoyed 59 drug years of prior availability. Of the other 98 drugs that were not mutually available, 73 were exclusively available in Britain, while 21 were exclusively available in the United States. . . . Many of the drugs exclusively available in Great Britain had become the drugs of choice in British teaching hospitals.

Significantly, of those drugs exclusively available in Great Britain, only two were eventually withdrawn from use because of toxicity, and these contributed negligibly to total drug toxicity in Great Britain during the study period.[35]

The way the two nations have chosen to address the risks associated with the licensing of ethical drugs closely parallels their approach to regulating potential hazards in other dimensions of health and safety. While the British system, by expediting the approval of new drugs, makes it more likely that its citizens will be harmed by too little regulation, it minimizes the likelihood that restrictions will be imposed that do not contribute to the health and safety of the public. The American strategy produces precisely the opposite result: it makes it less likely that its citizens will be exposed to hazards that could have been avoided through stricter controls but more likely that costs will be imposed on industry that produce no discernible benefits to the public. Much the same could be said of the two nations' approaches to risk assessment in the area of environmental regulation. Thus in all three areas—occupational health and safety, environmental protection, and consumer protection—the approach of American authorities to risk assessment has been more conservative; British officials consistently appear willing to take somewhat greater risks. Yet at the end of the day, as the English say, the American public has received no better protection; indeed, in the case of drug regulation it actually appears to be somewhat worse off.

These comparisons of health and safety regulation in areas other than environmental protection also suggest that the pattern of ownership is *not* a critical factor in accounting for the differing patterns of government intervention in the two societies. For while some major industrial sources of air and water pollution in Great Britain—most notably automobiles, steel, and utilities—are either wholly or predominantly owned by the British government, the industries whose regulation we have reviewed in detail are not. Asbestos, vinyl chloride, pesticides, chemicals, medicines, food and food additives are all exclusively manufactured by privately owned firms in both countries. And yet the differences we have observed in the area of pollution control persist. Moreover, if the pattern of ownership were of political significance, one would expect the dynamics of regulation to have been affected by the significant expansion of the British public sector over the last half-century. But this has not been the case: the British approach to social regulation has remained remarkably stable over the last century. On the other hand, the relationship between publicly owned utilities such as TVA and regulatory authorities in the United States has been no less contentious than that between investor-owned utilities and the EPA.

## CONSUMER PROTECTION

Like the United States, Britain has, over the last two decades, enacted several new laws designed to increase the amount of protection available to consumers.[36] The Consumer Protection Act (1961), the Weights and Measures Act (1963), the Trade Descriptions Act (1968), the Fair Trading Act (1973), and the Consumer Credit Act (1974) regulate both product description and product quality and are roughly analogous in scope to statutes enacted over a similar time period by the United States government. They differ, however, in two important respects. First, while U.S. consumer protection laws tend to deal with either specific kinds of products (i.e., automobiles) or hazards (i.e., radiation), Britain's laws have been much more general. As a result, Congress has enacted far more consumer legislation than Parliament.

Second, a relatively larger proportion of consumer protection legislation is administered by the national government in the United States than in Britain, although the role of the states in this area remains extremely important. In Britain the enforcement of consumer protection legislation has remained almost exclusively in the hands of local authorities. The two most notable exceptions are drugs, which are regulated by the Department of Health and Social Security, and agricultural products, which are under the jurisdiction of the Ministry of Agriculture, Fisheries, and Food. In 1973 an Office of Fair Trading was established to encourage competition and protect consumers, but in the main the Department of Prices and Consumer Protection was given little direct regulatory authority; its responsibilities lay primarily in supervising and coordinating the efforts of local consumer agencies. (It was subsequently abolished in 1979.) Unlike the United States' new consumer protection laws, then, Britain's have not been administered by newly established regulatory bodies; rather, in most cases the scope of the authority of existing regulatory units has simply been expanded.

During the 1960s and 1970s many American regulatory officials at the federal level came to identify their mission as that of protecting the interests of consumers vis-à-vis business. Indeed, several officials appointed to regulatory agencies during the Carter administration had previously been active as consumer lobbyists. In Britain, however, the officials of consumer agencies continue to define their roles as those not of advocates but of impartial arbiters. One British enforcement official noted:

> We don't see ourselves as the British version of Nader's Raiders. But we do see ourselves as standard bearers—trading standard bearers—committed to the continuing task of ensuring those standards are respected and neither flouted nor eroded. Unlike Mr. Nader we don't see ourselves wholly involved in taking sides in a fight. We know from experience that we are not best employed fighting battles for or with consumers.[37]

On the whole, consumer agencies in Britain enjoy extremely good relationships with the companies whose conduct they are responsible for regulating: "enforcement officials see themselves not as policemen whose sole duty is to detect offences, but as officials who must prevent them before they occur."[38] Moreover, they appear to have a higher opinion of the integrity of the companies with which they interact than at least some of their counterparts in the United States.

As in the case of pollution control, the public portrayal of business as basically law-abiding and well intentioned is critical to the enforcement strategy of British consumer protection agencies.

> If business were publicly labelled as unscrupulous or exploitative, its resistance to regulation would increase. By emphasizing that business is mainly law-abiding, the state and its enforcement agencies can reduce the fear of regulation and maximize the potentiality of obtaining compliance with their designs through persuasion.[39]

In America, by contrast, the enforcement of many consumer protection laws became more adversarial during the 1970s, particularly at the federal level, with government officials frequently publicly chastising companies for their insensitivity to the welfare of consumers and ordering a large number of product recalls. American executives have a long litany of cases of "unreasonable" enforcement on the part of such agencies as the Federal Trade Commission, the National Highway Traffic Safety Administration, the Consumer Products Safety Commission, and the Food and Drug Administration.[40] In Britain, industry and government have experienced far less conflict in this policy area. Unfortunately, we lack any data as to the relative effectiveness of the two nations' regulatory styles.

## ECONOMIC REGULATION

There is little point in comparing the patterns of economic regulation in the United States and Great Britain since their institutional structures vary so enormously: industries that are regulated in the United States, such as airlines, telephone and telegraph companies, railroads, and utilities, have generally been publicly owned in Britain. This is not the case, however, in three important sectors: banking, securities, and insurance. With the exception of the Bank of England, which does not engage in direct commercial lending, all financial institutions are privately owned in both countries. How does each country go about the regulation of these institutions? The two nations' approaches to the regulation of these sectors closely parallel the patterns we have observed in the areas of health and safety regulation and consumer protection. While the British make extensive use of self-regulation and place considerable reliance on informal social controls, the American approach is rule-oriented and legalistic. On balance, the amalgam of British financial institutions that constitute "the City" enjoy less nonstatutory regulation than any other major financial center; American financial institutions and markets are more strictly regulated than those in any other capitalist country.[41]

The American securities industry is formally regulated by the Securities and Exchange Commission (SEC), established in 1934 in response to the scandals associated with the stock market collapse of 1929. While relations between the SEC and the industry it regulates have in general been more cooperative than has been the case with such social regulatory agencies as the EPA, OSHA, and the FDA, its basic approach to regulation is similar: it promulgates a large number of detailed and specific rules, issued after a set of formal hearings, and frequently resorts to prosecution when these rules have been violated. While Congress has tended to place fewer restrictions on the SEC than on many other regulatory agencies, the agency's decisions are subject to judicial scrutiny. Each of the major stock exchanges has its own governing committee that establishes rules for the firms under its jurisdiction, but ultimate authority rests with the SEC.

Britain has enacted a number of statutes establishing regulations for publicly held companies. These regulations are extremely detailed and roughly analogous to those promulgated by the Securities and Exchange Commission. The first British Company Act was enacted in 1948 and the most recent in 1981; the latter implemented the requirements of the European Community's fourth directive on company accounting and reports: its "provisions on the form and content of the accounts and the notes to the accounts represent some of the most detailed and prescriptive requirements in British common law."[42] However, the British government plays virtually no formal role in the regulation of securities markets themselves. Whatever regulation exists takes place primarily under the auspices of two private organizations: the Council for the Securities Industry, established in 1978, and the Panel on Takeovers. The former is composed of a number of powerful bodies within the City, including the Stock Exchange Council, which oversees listed and unlisted companies and their dealings on the exchanges, while the latter is responsible for enforcing a voluntary code on takeovers and for policing the stock market on a deal-by-deal basis. Neither body enjoys any legal authority. There does exist a Prevention of Fraud (Investments) Act and the Department of Trade did issue new regulations following the folding of a major licensed securities dealer in 1981, but their scope is limited.

Banks in the United States operate under a complex set of federal and state regulations, the most important of which originated in the 1930s. Until the late 1970s, however, banks in Britain were free from statutory controls. In 1979, Parliament, partially in response to an EEC directive requiring each member state to establish a uniform system of banking regulation, enacted the 1979 Banking Act. This law, however, established only a deposit protection scheme and a two-tier structure of recognized banks and licensed deposit-taking institutions. All other regulatory functions are still performed by the Bank of England, which continues to be responsible for approving applications for new banks, judging them on such vague criteria as the applicants' reputations and standing in the financial market. The Bank also monitors the activities of banks by examining their quarterly returns and conducting regular interviews with senior managers. Instead of applying fixed rules, it attempts to influence key individuals. One observer notes: "Frequent discussion between senior management of banks and senior officials of the Bank of England are more conducive to the maintenance of good banking practices than the technique adopted in many other countries of sending in teams of inspectors to examine the banks' books." The style of these discussions, which in the Bank's words are "the cornerstone of the ... system of supervision," is "designed to recreate the intimacy and informality of the old approach to regulation."[43]

The two nations' approaches to insurance regulation follow similar patterns. In the United States, the industry is supervised by regulatory commissions established by the various states. While theoretically the industry in Britain is supervised by the Department of Trade, in practice this extremely important British industry is regulated by a body known as the Council of Lloyd's. Most of its twenty-eight members are drawn from the industry itself. Following scandals in 1979 and 1980, Parliament enacted the Lloyd's Act, which made it considerably easier for the council to exercise disciplinary powers over its member companies. And in 1982, following another series of scandals—the most dramatic of which involved an American company—that "raised doubts about Lloyd's ability to regulate questionable transactions by its brokers and underwriter members," Parliament approved additional controls.[44] Under the terms of the statute enacted in 1982, brokers who do business with Lloyd's are forbidden to own a controlling interest in a Lloyd's underwriter; this separation must be complete by 1987. In addition, the public will now be provided with substantially more information about the financial activities of Lloyd's brokers. Lloyd's itself, however, will continue to enjoy substantial independence from direct government supervision.

As in the case of ethical drugs, the regulation of British financial institutions and markets has become more detailed and legalistic over the last decade. To a significant extent Britain has moved from a system based on self-regulation to one that combines self-regulation with a more formal system of statutory controls. Indeed, no area of British regulation of industry has been in such a state of flux since the late 1970s. While in the past the British sense of "fair play" made a system of regulation essentially based on clubbishness a viable one, its effectiveness has been increasingly called in question, in part as a result of the growing number of non-British firms with offices in London. Yet while the gap between British and American standards and policies has narrowed, substantial differences persist. In general, American regulations are stricter, are more formal, and provide much less scope for self-regulation. Britain places fewer restrictions on insider trading than does the United States, for example, and while the Securities and Exchange Commission closely monitors—and on occasion has overruled—the accounting standards established by the Financial Accounting Standards Board, in Britain the Accounting Standards Committee, composed of Britain's six accounting bodies, both determines standards and is responsible for enforcing them. In the related area of antitrust policy, Britain's Monopolies and Mergers Commission approaches each takeover or merger on an ad hoc basis. In America, in contrast, antitrust policy proceeds through the application of explicit standards whose interpretation is then subject to exhaustive judicial review.

The far stricter legal controls exercised by American regulatory authorities appear to have been no more successful in preventing periodic scandal and frauds; such episodes continue to occur in both societies. On balance, British holders of insurance policies, bank depositors, and investors appear to be no better or worse protected than their counterparts in the United States.

This brief survey of economic regulation also reinforces the argument that state ownership in Great Britain is not a critical factor in accounting for the differences in the regulatory policies pursued in the two countries. In these three sectors, the patterns of ownership in the two societies are identical, yet the differences in the styles of regulation that we have observed in other areas persist. It also suggests the limits of purely economic factors in accounting for the differences in the patterns of regulation in the two societies. For while many British industries, public as well as private, that directly affect public health and safety have certainly been less profitable than their counterparts in the United States, British financial institutions have been, and remain, extremely competitive in the world marketplace. (The same can be said of the British pharmaceutical industry.) And yet the British and American regulatory policies in these sectors show the familiar contrasting patterns.

## CONCLUSIONS

In sum, each nation does exhibit a distinctive regulatory style, one that transcends any given policy area. The British government regulates the impact of business decisions on the environment in much the same way it attempts to control a variety of dimensions of corporate conduct. Regulation of industry tends to be more informal in Britain than in the United States, more flexible, and more private. Regulatory officials are able to exercise considerable discretion and tend to make policy on a case-by-case basis rather than through the application of general rules and standards. Little use is made of prosecution and much reliance is placed on securing compliance through informal mechanisms of social control, including, in many instances, self-regulation. Regulatory officials tend to have close working relationships with the members of those industries whose conduct they are responsible for supervising; the latter are closely consulted before rules are issued and regulations enforced. In America, on the other hand, regulation tends to be highly formalized: it proceeds on the basis of the application of broad rules that are made and enforced in accordance with strictly defined procedures. The entire regulatory process is subject to close scrutiny by the courts, the legislature, and the public as a whole. Fines are levied for violations relatively frequently and little reliance is placed on industry self-regulation.

Participation by nonindustry constituencies does vary widely across regulatory areas in both countries. The Alkali Inspectorate and the system of land-use planning, for example, represent virtually polar opposites in terms of the opportunities they extend for participation by nonindustry constituencies. On the other hand, the American system of regulation does not invariably provide more opportunity for nonindustry participation than the British: trade unions are more directly involved in the making and enforcement of occupational health and safety regulation in Britain than in the United States. The critical distinction between the British and American approaches to regulation has to do with the terms in which nonindustry participation takes place.

In America there are numerous points of access to the regulatory process: interest groups can lobby Congress, participate in agency rule-making procedures, attempt to influence the regulatory review process at the White House, and challenge particular regulations in the courts. If state and local governmental units are also affected, the opportunities for intervention are multiplied still further. In Britain the number of forums at which the public can intervene in the regulatory process is more limited; to the extent that public participation takes place at all, it is generally restricted to a handful of arenas. An important reason why government regulation of industry has created so much less overt political conflict in Britain than in the United States is that the relationship among the interest groups involved in particular policy areas tends to be relatively structured. This is clearly true in the case of industry, whose interaction with government officials usually takes place through trade associations, though large companies generally negotiate with the government directly. But to a significant extent it is also true of nonbusiness constituencies: both by controlling access and by using private organizations to implement various regulatory policies, the British government has played an important role in shaping both the number and scope of the political pressures placed upon it. Alternatively, it is in part the relative openness of the American regulatory process that makes it so contentious.

There are exceptions to each of these generalizations. The making of British land-use policies takes place in highly visible forums and the procedures governing the "large public inquiries" are highly legalistic. In addition, over the last decade the regulation of both ethical drugs and the City has come to resemble more closely the more formal approach adopted earlier by the United States. Nor are the politics of British regulation invariably less contentious: the four case studies of British environmental policy described in Chapter 1 demonstrate the extent to which disputes over various policies have spilled over from the forums originally established to contain them and become highly public, involving both Parliament and the cabinet. Yet if one compares policies in any particular area of regulation, significant differences remain. Particularly over the last fifteen years, a significantly larger proportion of regulatory policies have been formulated and implemented through relatively informal and private negotiations between regulatory officials and interest groups in Britain than in the United States. On the whole, the regulatory process has been far more public—and therefore more politicized—in the United States than in Great Britain. It has not, however, been notably more effective.

This conclusion does not, however, unequivocally support the position of political scientists who contend that policy process and outcomes are primarily the result of factors peculiar to the countries where they are developed, that is, that "politics matters." For the agenda of regulatory policy itself does appear to be shaped by socioeconomic factors that transcend any particular political system. It is not merely that the issues of environmental quality and of worker and consumer health and safety in general have become more salient in both Great Britain and the United States over the last two decades. Rather the specific ways in which these issues have been defined are strikingly similar. Consider the following examples of issue convergence: the construction and expansion of airport facilities, highways, and offshore energy facilities (i.e., LNG), the safety of nuclear power (Windscale and Three Mile Island), the health impact of exhaust from the internal combustion engine (i.e., lead), the effects of pesticide use on the health of both birds and agricultural workers (DDT, dieldrin, 2,4,5-T), the environmental impact of energy production (the North Sea and the North Slope), the pollution produced by the burning of coal in power plants (acid rain), offshore oil spills (Santa Barbara and *Torrey Canyon*), the disposal of toxic wastes (Love Canal, Nuneaton), health hazards to workers (asbestos and vinyl chloride), the safety of ethical drugs (thalidomide, Oraflex).

This convergence of regulatory agendas appears to be due to two factors. One is the similar levels of industrial development in the two nations.[45] The economies of both Britain and the United States are dominated by many of the same industries—indeed, in many cases, the very same multinational companies. Obviously these firms—and their products—produce many of the same externalities. Moreover, because significant segments of their populations are both affluent and highly educated, they have similar expectations: they are in a position to demand—and to pay for—less hazardous working conditions, safer products, and a more scenic and healthful physical environment. Industrial growth in a democratic society thus simultaneously creates environmental problems and places pressures on policy makers to ameliorate them. It is certainly not coincidental that public interest in these policy areas increased in both countries in the late 1960s and early 1970s, following two decades of relatively sustained and rapid economic growth.

There is a second explanation for the convergence of the regulatory agendas of Great Britain and the United States in the postwar period: international communication. The media have played an important role in disseminating information about various regulatory issues to interested citizens on both sides of the Atlantic. The London "killer fog" of 1952, the blowout at Santa Barbara in 1966, the wreck of the *Torrey Canyon* in 1965, the publication of *Silent Spring* in 1962, Love Canal in 1977—all received considerable press coverage in both countries. Policy makers and scientists have also communicated extensively: they attend the same conferences, read the same journals, and regularly exchange technical information. While on occasion scientists in Great Britain and the United States disagree about the hazards associated with particular products, chemicals, or technologies, these differences are far outweighed by their areas of agreement. This scientific consensus has played an important role in shaping the regulatory agendas of both nations. As a result, whenever one nation has identified a particular product, chemical, or production process as hazardous, officials on the other side of the Atlantic invariably find themselves under pressure from both activists and scientists in their own country to do likewise.

Since the early 1970s, activists in both countries have developed increasingly close ties: both Greenpeace and Friends of the Earth have British and American chapters, and Ralph Nader provided some of the initial funding for the investigatory journalism of the *Social Audit*. The German Marshall Fund regularly sends American environmentalists to European countries, including Britain. British environmental organizations, with fewer resources and with less access to government documents, have come to rely on their counterparts in the United States for information, and so raise in Britain many of the issues that have recently appeared on the political agenda in the United States.

Why do regulatory outcomes appear relatively similar in the two countries? Why have they been only marginally affected by the substantial differences in the ways in which policies have been made and enforced? One reason is that policy makers in both capitalist democracies operate under a similar set of constraints. For while economic growth both creates externalities and provides the available resources to ameliorate them, it also constrains the amount of resources that can be committed to such efforts. Obviously, national priorities do differ somewhat and there can be legitimate disagreements about what a nation or a particular industry, firm, or plant can "afford." But in the long run, the severity of enforcement is strongly influenced by the interests of policy makers, industrial workers, and the public as a whole in keeping their nation's industries internationally competitive. Again, it is not coincidental that enforcement efforts in both nations slackened somewhat following the increase in oil prices in late 1973. Moreover, as I have previously argued, regulation is only one factor among many that affect either environmental quality or public health and safety. To the extent that the actual quality of the environment varies in Great Britain and the United States, the difference appears to be due less to their systems of regulation than to geographical conditions and industrial and technical factors.

Yet at the same time, the ways in which regulatory decisions are made vary substantially in Great Britain and the United States. In this context, the importance of each nation's political system remains decisive: these are distinctive national styles of regulation. Compared to both the regulatory agenda and regulatory outcomes, the way in which each political system has gone about making and implementing regulatory controls remains highly distinctive. The latter dimension appears to have been much less affected by technological or socioeconomic factors than the former two. The evidence presented in these two chapters does not of course resolve the debate over the relative importance of political and structural factors in shaping public policies across national boundaries. Their conclusion is valid only for the two countries and the limited numbers of issues they have addressed. At the same time, the body of scholarly literature on this subject does suggest that these generalizations can be applied to other industrialized democracies as well.

Lennart Lundquist's *Hare and the Tortoise* compares the regulation of air pollution in the United States and Sweden during the 1960s and 1970s.[46] In two critical respects, Swedish and British public policies in this area are roughly comparable: both nations have placed primary emphasis on the reduction of pollution from stationary sources and both have tailored their regulatory requirements to the capacity of industry to comply with them. When the number of miles traveled is taken into consideration, automobile emissions did decline more rapidly in the United States than in Sweden. Because the Swedes made more progress in reducing pollution from stationary sources, however, the overall levels of air pollution in the two countries declined to a virtually identical degree: approximately 8 percent in Sweden and 10 percent in the United States. This similarity in regulatory outcomes is particularly striking when one reflects that Sweden is one of the most environmentally conscious nations in Europe. Moreover, the control of air pollution produced far less conflict there than in the United States.

In 1982 an exhaustive and comprehensive cross-national study of environmental regulation was completed. Conducted under the auspices of the International Institute for Environment and Society in Berlin, it examined the implementation of controls over the emissions of sulfur dioxide in ten Western European countries. The authors of the study found that there was no systematic relationship between the formal regulatory system adopted by each country and its actual level of sulfur dioxide emissions; in fact, the levels of sulfur dioxide emissions actually varied more *within* each country than among countries. They found no relationship between the amount of public participation in the regulatory process permitted in each country and the stringency of the permits issued by its regulatory authorities. The project's directors conclude:

> The output analysis meticulously carried out in . . . selected [regions] . . . did not reveal any significant differences with regard to the environmental substance of the consents between those countries with a more open enforcement process (e.g., FRG, Belgium, France, and the Netherlands) and those with closed administrative procedures; i.e., those without institutionalized inclusion of members of the affected public (e.g., Italy, Switzerland, and the United Kingdom).[47]

The conclusions of both these studies are similar to those reached by three other cross-national studies of social regulation. The study of chemical regulation in Great Britain, France, the United States, and the Federal Republic of Germany carried out by Ronald Brickman and his colleagues and Joseph Badaracco's analysis of the control of vinyl chloride in the United States, France, Great Britain, and Japan—both of which have been summarized here—make the same argument: regulatory agendas and outcomes vary much less than regulatory processes.[48] While Steven Kelman's book *Regulating America, Regulating Sweden* does not explicitly deal with the issue of effectiveness, it draws a similar conclusion: occupational health and safety regulations are comparable in the two countries but the level of political conflict is substantially greater in the United States.[49]

These studies and my own analysis suggest that at least among the advanced capitalist democracies, the politics and administration of amenity and health and safety regulation vary much more than either actual regulatory outcomes or the political agenda itself. We now need to account for the differences in regulatory style between Great Britain and the United States; that is the subject of Chapter 6.

# The Dynamics of Business–Government Relations in Great Britain and the United States

Why have the United States and Great Britain evolved such different approaches to the regulation of industry, not only in the area of environmental regulation but in a whole host of other dimensions of regulatory policy? The two nations' regulatory styles were not always so dissimilar. In fact, many of the most critical differences in British and American regulatory practices have arisen only in the last two decades. Throughout the nineteenth century, government regulation of corporate social conduct followed similar patterns in the two countries. During the period of rapid industrial growth—between 1750 and 1850 in Britain, during the latter third of the nineteenth century in the United States—the relations between business and government were relatively adversarial in both societies. They subsequently became significantly more cooperative during the period of mid-Victorian reform in Britain (roughly 1860–1880) and the Progressive Era in the United States (1902–1916). In Britain this cooperative relationship has persisted through the present day; in America it atrophied. Confronted by roughly similar sets of policy issues during the late 1960s and early 1970s, policy makers in Britain addressed them in much the same manner as they had addressed such issues a century earlier; in America the making and enforcement of regulatory policy became transformed. Why?

### THE ADVERSARY SYSTEM

The role played by government in shaping the economic development of both Britain and the United States, the first two nations to industrialize, was a modest one.[1] In both countries capital was accumulated and allocated primarily through the market, with entrepreneurs rather than government officials playing the most critical roles. Not surprisingly, the British and American business classes developed virtually identical sets of ideologies that both reflected and reinforced the notion that government could promote economic growth most effectively by interfering with business as little as possible. As Herbert Spencer, the English social theorist whose views became highly influential in the United States, wrote, "badly as government discharges its true duties, any other duties committed to it are likely to be still worse discharged. To . . . regulate, directly or indirectly, the personal actions of [citizens] is an infinitely complicated matter. . . . Unless spontaneously fulfilled, a public want should not be fulfilled at all."[2]

The principles of paternalism that had informed upper-class thinking in Great Britain eroded rapidly, even before Britain had become a full-fledged industrial society. Twenty years after the publication of Adam Smith's *Wealth of Nations*, William Pitt, the British prime minister, summarized the prevailing consensus in the course of a debate over whether the government should regulate wages in order to prevent the laboring poor from starving. Pitt informed Parliament: "It was indeed the most absurd bigotry, in asserting the general principle [of political economy], to exclude the exception; but trade, industry and barter would always find their own level, and be impeded by regulations which violated their natural operation and deranged their proper effect."[3] The historian Oliver MacDonagh adds:

> . . . by and large it was accepted dogma [among enfranchised Englishmen during the middle decades of the nineteenth century] that competition, material self-interest and the profit-motive should be given as wide a field for action as was compatible with the bare necessities of enforced contract and public order. As to efficiency, celerity, and economy, individual and governmental activity were simply opposites.[4]

Accounts of nineteenth-century business thinking by such American historians as Richard Hofstadter and Robert McClosky reveal a similar outlook.[5]

During this period the ideal of the self-made man became highly influential in both societies. Echoing the theme of the Horatio Alger novels that were widely read in the United States, the Englishman Samuel Smiles wrote in his book *Self-Help*, "What some men are, all without difficulty might be. Employ the same means and the same result will follow."[6] To compare the novels of Charles Dickens with those of Upton Sinclair or

Theodore Dreiser or the exposés of the muckrakers with the reports of the various royal commissions on factory conditions (immortalized in the footnotes of volume 1 of Karl Marx's *Capital*) is scarcely to be struck by any significant differences in the behavior or social and political outlooks of industrialists in the two countries.

In both societies the entrepreneurial ideal was, in principle, strongly opposed to government intervention designed to ameliorate the social and physical impacts of rapid industrial growth. Charles Dickens, with his usual eloquence, evokes this world view in his novel *Hard Times*:

> Surely there was never such fragile china ware as that of which the manufacturers of Coketown were made. Handle them ever so lightly and they fell to pieces with such ease that you might suspect them of having been flawed before. They were ruined when they were required to send labouring children to school; they were ruined when inspectors considered it doubtful whether they were quite justified in chopping people with their machinery; *they were utterly undone when it was hinted that perhaps they need not always make quite so much smoke.*[7]

Commenting on the British government's first efforts to regulate working conditions, Marx wrote in *Capital*:

> If it were made obligatory to provide the proper space for each workman in every workshop...the very root of the capitalist mode of production ...would be attacked. Factory legislation is therefore brought to deadlock before those five hundred cubic feet of space.... The Factory inspectors ...harp over and over again upon the necessity of those five hundred feet and upon the impossibility of wringing them out of capital.[8]

In his history of early Victorian government, Oliver MacDonagh writes:

> By and large as a group the manufacturers and business interests fought each attempt at interference as if it were a matter of life and death. So indeed they persuaded themselves it was. It was no caricature when their opponents depicted them as arguing that no profit whatever was made except in the last hour of work, or that the prosperity of England rested upon the labours of half a million little children.[9]

One of the early British factory acts was criticized by a Scottish manufacturer as "indefensible in principle; invidious, oppressive, and absurd in its provisions; in its penalties harsh, ruinous, and tyrannical in the extreme." A student of British regulatory policy adds that "the early British legislation on workplace health and safety, and the inspectors who enforced it, met with strong and sometimes bitter opposition from both employers and political leaders."[10] The efforts of various American state governments during the latter part of the nineteenth century to protect the interests of industrial workers, farmers, and on occasion consumers likewise met with strong and bitter resistance from business.

The limited role played by government in either country in the nineteenth century need not be exaggerated.[11] Neither political system came even close to approximating the ideals of laissez-faire. The Englishman James Bryce wrote in his study of American government, *The American Commonwealth*, published in the 1880s:

> Though the Americans have no theory of the State and take a narrow view of its functions, though they conceive themselves to be devoted to *laissez-faire* in principle, and to be in practice the most self-reliant of peoples, they have grown no less accustomed than the English to carry the action of government into ever widening fields.[12]

On balance, the American government, particularly at the local level, played a more active role in promoting economic development.[13] Andrew Shonfield writes that "the characteristic role of government in the United States during the first half of the nineteenth century was its emergence as an entrepreneur on a large scale."[14] While virtually all transportation facilities in Britain were privately owned and financed, for example, American state and local governments played a critical role in both supplying their capital and managing them. On the other hand, the British government tended to exercise much greater control over their construction: "The enactment of general railway laws in the United States, throwing open the power of constructing railroads at every and any point, ... was a condition utterly unknown and remains unknown to the English people."[15]

The earliest form of government regulation in both societies was exercised by the courts. In the area of pollution control, for example, people whose property rights were adversely affected by the emissions of their neighbors had the right to claim compensation under the common law doctrine of nuisance—a right that Britain's first pollution-control statute, enacted in 1821, sought to expand. This avenue, however, proved of limited effectiveness in both countries, largely because American courts had adopted the British "standard of care" doctrine. As one legal historian writes, "the legislatures, the courts, and certain segments of the public [favored] industrialization, and they were not anxious to burden industry with damage actions."[16] As a result, it became relatively difficult to collect damages from polluters on the basis of the common law doctrine of nuisance in both countries. Because a larger proportion of British streams and rivers were privately owned, however, the courts were more willing to award damages for water pollution in Britain than in the United States.

Governments at the national as well as the local level in both countries did make some efforts to safeguard public health, establish safety regulations for novel and unusually dangerous technologies, and temper some of the abuses associated with the employment of women and children.[17] These measures tended to be strongly opposed, however, and often were significantly weakened as a result of political pressure from industrialists; moreover, only the most rudimentary administrative machinery existed to administer and enforce them. They thus amounted to little more than ad hoc accommodations to particularly blatant abuses.

From the very outset, British regulatory officials adopted a relatively conciliatory stance toward industry, relying on persuasion and negotiation rather than prosecution. This policy was dictated largely by political expediency: given the initial hostility of the manufacturing interests to the enforcement of the factory laws and the meager administrative resources available to the inspectorates established to enforce them, they had little choice but to seek to secure voluntary compliance.[18] Much the same was true in the United States. Charles Francis Adams, who headed the nation's first important regulatory agency, the Massachusetts Board of Railroad Commissioners, took considerable pains to avoid antagonizing the railroad industry. Shunning prosecution, he emphasized voluntary compliance and extensive consultation with industry officials.[19]

Ironically, while the regulatory statutes enacted by the states in the nineteenth century tended to be so vague that administrators had considerable flexibility in interpreting them, some of the earliest regulatory statutes enacted in Great Britain tended to minimize the exercise of administrative discretion, in part as a way of ensuring at least some rudimentary level of enforcement. The 1844 Factory Act, for example, represented "a marked reduction in the discretion left to the inspectors, ... in the assessment of the age of children and in the standards for schools."[20] Similarly, the first alkali acts contained a specific emission standard for hydrochloric acid, and the air-pollution control legislation enacted in 1881 established fixed, statutory standards for sulfates and nitrous oxides. (It was only toward the end of the nineteenth century that enforcement began to be based on the more flexible concept of "best practicable means," and it was not until 1906 that the placing of a new pollutant under the jurisdiction of the Alkali Inspectorate no longer required an act of Parliament.) In addition, Parliament was far more actively involved in approving legal changes in land use during the eighteenth and nineteenth centuries than it has been in the twentieth; each enclosure, for example, required parliamentary approval. Even the number of prosecutions of companies for violating the factory acts appears to have been somewhat greater in the first than in the second half of the nineteenth century.[21]

## THE MOVE TOWARD REFORM

Both societies subsequently experienced a major shift in the climate of business and intellectual thinking—one that had important consequences for the development of government regulation. Parallels can be seen between the mid-Victorian reforms of the 1860s and 1870s and those of the Progressive Era, a generation later. Most important, both reform movements drew their primary support from the same social base: "the non-capitalist or professional class."[22] Composed of neither businessmen nor workers, this new middle class had rapidly developed in both Great Britain and America during the latter decades of the nineteenth century. Strongly influenced by the doctrines of evangelical Protestantism in both countries (a disproportionate number of Victorian reformers were nonconformists and many Progressives were the children of Protestant ministers), many of its more articulate members became highly committed to the cause of political and social reform. These "ministers of reform" approached the amelioration of sufferings and injustices produced by rapid industrial development with a sense of "evangelical dedication."[23] (It is not coincidental that both the Progressives and the Victorian reformers accorded a similar priority to the curbing of prostitution.) These reformers saw themselves as playing a mediating role in the ongoing struggles between the manufacturers and the working class. Uncomfortable with the militancy of the latter and morally repelled by the shortsighted greed of the former, they promoted the values and ideals of professionalism, scientific and technical expertise, administrative competence and neutrality, and efficiency in both business and government. These individuals played a leading role in the effort to establish a professional civil service in both countries and they became its first administrators. Under their influence, the main mechanism for controlling corporate social conduct switched from the courts to administrative agencies.

Dr. George Wiley, who headed the Food and Drug Administration for more than a generation, and Gifford Pinchot, who helped shape conservation policy during Theodore Roosevelt's administration, represent the American equivalents of such Victorian civil servants as Edwin

Chadwick, who pioneered the development of the British welfare state; Angus Smith, the first chief inspector of the Alkali Inspectorate; and Alexander Redgrave, who served for thirteen years as chief of the Factory Inspectorate. On the whole, their approach to regulation deemphasized prosecution and encouraged cooperation; they saw themselves less as adversaries of industry than as technical advisers to it. Indeed, the dominant theme of Victorian administrative philosophy was "a belief in persuasion and knowledge as the best administrative devices."[24] Thus the Factory Inspectorate "came to see itself, somewhat self-consciously, as a body of public servants standing above class conflict."[25] The ideology of many regulatory officials during the Progressive Era was roughly similar.

The research and writing of this new middle class helped to transform public attitudes toward government intervention. In both countries the factor that precipitated each new expansion of government's role was the public exposure of a particular intolerable "social evil." In the United States these exposures were primarily the work of journalists, while in Great Britain an analogous role was performed by the investigations of government officials themselves.[26]

> In all publications,... whether newspaper, review, or pamphlet, the inspectors' reports opened the eyes of the decent and comfortable to the misery, ignorance, filth, crime, and degradation that were the lot of the working classes. The inspectors urged upon the upper classes the adoption of remedies which would lessen these moral and physical evils. Their reports made Victorian society more conscious of its failings, and inspired administrative actions which would remove the evils.[27]

Intellectuals in both countries helped furnish the theoretical justification for a significant increase in government regulation. Thus while John Stuart Mill wrote in 1848 that "letting alone...should be the general practice," two decades later he concluded: "In particular circumstances of a given age or nation, there is scarcely anything, really important to the general interest, which it may not be desirable or even necessary, that the government should take upon itself, not because private individuals cannot effectively perform it, but because they will not."[28] Echoing the later Mill, Herbert Croly wrote in *The Promise of American Life*, published in 1901:

> Reform is both meaningless and powerless unless the Jeffersonian principle of non-interference is abandoned. The experience of the last generation plainly shows that the American economic and social system cannot be allowed to take care of itself, and that the automatic harmony of the

individual and the public interest, which is the essence of the Jeffersonian democratic creed, has proved to be an illusion. Interference with the natural course of individual and popular action there must be in the public interest; and such interference must at least be sufficient to accomplish its purpose.[29]

In a similar spirit, the Cobden prize essay at Oxford in 1880 declared, "We have had too much of laissez-faire.... We need a great deal more paternal government—the bugbear of the old economists."[30]

Whether or not the doctrines of laissez-faire can be legitimately attributed to the influence of Jeremy Bentham remains a subject of considerable controversy among historians.[31] It is beyond dispute, however, that while Bentham's utilitarian ideas served to limit government intervention during the first two-thirds of the nineteenth century, by the 1860s they were being invoked to justify its expansion. As the English legal historian A. V. Dicey noted in his classic *Law and Public Opinion*,

> somewhere between 1868 and 1900... changes took place which brought into prominence the authoritative side of Benthamite liberalism. Faith in laissez-faire suffered an eclipse; hence the principle of utility became an argument in favour, not of individual freedom, but of the absolutism of the State.... [The] English administrative mechanism was reformed and strengthened. The machinery was thus provided for the practical extension of the activity of the State.[32]

MacDonagh similarly observes a "catastrophic and very general collapse of political individualism in the last quarter of the nineteenth century."[33]

As a result, an important change took place in the political climate of government regulation in Great Britain during the 1860s and 1870s. Up through the 1860s, for example, textile mill owners felt unfairly singled out for mistrust and special legislation and lobbied for the repeal of the factory acts. During this decade Parliament not only refused to deregulate working conditions in textile mills but extended the factory acts to cover a large number of additional industries. By 1871 factory inspectors had access to practically every nondomestic industrial establishment in England. The 1860s and 1870s also witnessed the first effective legislation protecting the public from wholesale fraud and poisoning by the adulteration of food as well as the beginning of the central government's effort to prevent and control air pollution. Shortly after Benjamin Disraeli's victory in the 1874 election, Parliament further extended the factory acts, granted local authorities the right to regulate various kinds of air pollution, and passed Great Britain's first water-pollution statute. Responding to a series of exposés of food adulteration in a magazine published by the British Medical Association, which anticipated those made a quarter-century later by Upton Sinclair in *A Call to Reason*, it subsequently strengthened and extended the regulation of food quality.[34]

These changes in public policy and public opinion both affected and reflected a change in the thinking and behavior of the business community. In England an important vehicle for this change in outlook was evangelicalism. Harry Eckstein writes that "amongst the upper classes ...a crusading private humanitarianism, an almost incredible amount of philanthropy ... the Evangelical influence ... blocked the way for the entrenchment of the harsher *laissez-faire* doctrine."[35] At the municipal level, according to Donald Read,

> the Birmingham leadership group, mainly Nonconformist businessmen and ministers, developed a "civic gospel" to explain and justify their municipal activity.... In a sermon upon "The Evils and Uses of Rich Men," the Reverend R. W. Dale, Birmingham's leading Congregationalist minister, urged middle class big businessmen to give their time and their talents to municipal work, to feel "called of God" to ensure adequate provision of poor relief, good drainage and lighting, good schools, "harmless public amusement," and honest administration.[36]

The business elite of Birmingham, as well as of other major industrial cities, responded by substantially improving both the quantity and the quality of services provided by local government. They organized municipal gasworks, established sanitation systems, and built impressive municipal buildings.[37]

In addition to becoming more active in philanthropy and civic affairs at the local level, a growing number of manufacturers began to recognize that government regulation did not necessarily threaten their profits. The world's largest calico printer, for example, told the Social Science Association in 1864: "The Factory Acts were opposed by many of us as economically unsound and as an unjust interference with the rights of labor and capital. *They have been soundly beneficial.*" Contrasting the cooperation he now received from factory owners with the obstructions that had confronted his predecessor, a factory inspector noted in 1870 that "the whole country is now of one mind—labour should be moderate, work rooms and factories should be made healthy and the young should be educated."[38]

Roughly thirty years after the Factory Inspectorate was first established, the roles of its inspectors shifted from pure enforcement to a mix of technical assistance and law enforcement. The chief inspector in

1878 attributed this change to the "increasing complexity of the work and the changing attitude of industry." A legal historian, Bernice Martin, concludes that "a recognition of the long-term economic benefits [of compliance] formed an important, and in the second half of the century, an increasingly prominent part of [the] process, by which inspectors were able to persuade manufacturers that compliance was indeed in their economic self-interest." Writing in 1862, one of the inspectors observed that manufacturers were "no longer opponents of factory legislation," but were "now generally the foremost advocates of improvement."[39] A few years later, a textile manufacturer presented two papers to the National Association for the Promotion of Social Science in which he argued the political and economic case for the extension of the Factory Acts. Joseph Badaracco reports a comment by an official of the Health and Safety Executive: "We in Britain had our own problems with factory inspectors being met at the factory door by owners with shotguns. But we solved that problem in the '70s. I mean the 1870's."[40]

The American equivalent of this change in business thinking was reflected in part in the emergence of the doctrine of corporate social responsibility. George Perkins, the manager of U.S. Steel, wrote in 1908:

> The larger the corporation becomes, the greater become its responsibilities to the entire community. The corporations of the future must be those that are semi-public servants, serving the public, with ownership widespread among the public, and with labor so fairly and equitably treated that it will look upon its corporation as its friend.[41]

And by 1914, Walter Lippmann could note in his study of the Progressive movement, *Drift and Mastery*:

> The cultural basis of property is radically altered, however much the law may lag behind in recognizing the change.... The men connected with these essential properties cannot escape the fact that they are expected to act increasingly like public officials.... What they will learn is that it is no longer their business.... Big businessmen who are at all intelligent recognize this. They are talking more and more about their "responsibilities," their "stewardship."[42]

Historians as diverse as Samuel Hays, Gabriel Kolko, James Weinstein, and Robert Wiebe have shown that important segments of the business community supported and on occasion even initiated virtually every major reform enacted during the Progressive Era.[43] As in Great Britain, the reform efforts of businessmen were particularly important at the local level. In America they played a critical role in reducing patronage and making the delivery of municipal services more efficient. According to Weinstein, "the business community, represented largely by chambers of commerce, was the overwhelming force behind both commission and city-manager movements."[44] Moreover,

> the folklore of the business elite came by gradual transition to be the symbols of governmental reformers. Efficiency, system, orderliness, budgets, economy, saving, were all injected into the efforts of reformers who sought to remodel municipal government in terms of the great impersonality of corporate enterprise.[45]

At the national level, urban merchants, bankers, and brokers were also extremely influential in the drive for civil service reform: "at the height of the reform movement, businessmen...constituted about half of the membership of the civil service reform associations."[46] Dewey Grantham notes that "far from representing a monolithic obstacle in the path of reform, [businessmen] supported the passage of many progressive measures and, ironically, encouraged the undermining of laissez-faire and the drift toward political and administrative centralization."[47] Gabriel Kolko concurs: "The basic fact of the Progressive Era was the large area of consensus and unity among key business leaders and most political factions on the role of the federal government in the economy."[48]

The perennial debate among historians as to whether or not the reforms of the Progressive Era were dominated by business misses the significance of this unique period of American history. For the most part, Progressive reforms were not so much pro- or antibusiness as they were based on changing perceptions among many companies and industries as to where their self-interest lay. Rather than regarding every expansion of government as a threat to their autonomy and profits, an enlightened segment of the nation's industrial and financial leadership began to recognize that government had a useful role to play in conserving natural resources, improving public health, curbing deceptive trade practices, reducing pollution, compensating workers for injury on the job, and assuring the public of safe products. This is the essence of the doctrine of "corporate liberalism."

In fact, many of the government regulations of business established during the Progressive Era both promoted the interests of business and improved the public welfare. The most well-known example is the Meat Inspection Act of 1906, which simultaneously improved working conditions in the slaughterhouses, safeguarded the quality of meat for consumers, and increased the access of the meat processors to foreign markets.[49] The other major social reforms of the Progressive Era—including the Pure Food and Drug Act, the inauguration of a system of workers' compensation at the state level, the establishment of commissions to regulate public utilities, and efforts to conserve the nation's physical resources—can be understood in similar terms. Though often enacted despite severe reservations on the part of some businessmen, their value to the long-term economic growth and social stability of the business system was acknowledged by many others. An official of the California Industrial Accident Commission, charged with enforcing worker safety laws, stated in 1914:

> The attitude of the Safety Department toward employers and employees is not one of compulsion, but of cooperation. It is expected that compulsion will have to be resorted to in rare cases only. The letters on file show that the keenest interest is evinced by manufacturers who express in highest terms their appreciation of the practical suggestions offered by [our] safety engineers.[50]

It would certainly be misleading to describe the Progressive Era simply as one of cooperation between business and government. Rather the degree of consensus in regard to government's appropriate role in the regulation of corporate social conduct was higher in the first decade of the twentieth century than in other periods. In contrast to the conflict that attended both the New Deal and the "new wave" of social regulation enacted during the 1960s and 1970s, a disproportionate amount of the conflict over particular regulatory policies in the Progressive Era tended to occur within the business community; relatively few regulatory issues pitted business against nonbusiness constituencies.[51]

## THE CONTROL OF AIR POLLUTION

The earliest efforts of the British and American governments to control air pollution clearly illustrate the parallels between the early regulatory systems of the two countries. Toward the end of the nineteenth century, many American cities began to experience levels of air pollution similar to those that occurred in London; indeed, Pittsburgh newspapers described the periodic combinations of smoke and fog produced by the burning of high-sulfur coal that so dramatically reduced visibility as "Londoners." As in London and other English industrial cities, residents in cities throughout the American northeast began to press for restrictions on the burning of coal. By the early 1900s, most cities had their equivalents of London's Smoke Abatement Society. The British organization had to seek Parliament's approval to enact tighter regulations for the city of London; its American counterparts were required to petition their state legislatures before they could enact smoke ordinances. Both reform movements included among their supporters prominent businessmen who were themselves inconvenienced by the level of air pollution in the neighborhoods in which they lived and worked. Eventually the efforts of American reformers were successful, and by the middle of the Progressive Era most industrial cities had enacted smoke-control ordinances.

There was some initial controversy over the most effective strategy for securing compliance with these newly enacted laws. Some inspectors urged increased prosecution. The smoke inspector for Chicago wrote to his successor: "The way to abate smoke is to abate it. I have to suggest three remedies. The first is to fine the violators. The second is to fine them again. The third is to keep on fining them until they are bankrupt or repentant." This was a minority position, however, and its influence markedly declined in the second decade of the century. The view more commonly held was expressed by Pittsburgh's smoke inspector, who noted that the best smoke abatement was "up-to-date" smoke abatement, one in "harmony with the best spirits of the time." As "this is an age of education, cooperation, efficiency and economy," he said, he urged that the "strong arm and the Big Stick" approach be used sparingly and only against the "ignorant, selfish, lazy and wasteful." In 1911 a meeting of a large number of smoke-abatement organizations held in St. Louis passed a resolution declaring that "the smoke nuisance can be more quickly and effectively abated by education than by prosecution and that prosecution in the courts should only be instituted when cooperative methods fail."[52]

By the time of World War I, the Progressive notion that regulatory commissions should be staffed by experts had begun to affect the smoke-abatement movement; the various state and local regulatory agencies became increasingly staffed by engineers with experience in fuel economy. Their efforts became more technically oriented and regulators devoted considerable resources to educating managers about the most recent developments in smoke-abatement technology. In 1920 Pittsburgh's bureau chief boasted that "many of the men who were active opponents of [smoke abatement] . . . became warm supporters when they found out that compliance . . . saved them money."[53]

The early history of the Alkali Inspectorate followed a similar pattern.[54] By the middle of the nineteenth century Great Britain had established an important heavy chemical industry. In 1862 the processing of sodium carbonate to produce glass, soap, and textiles employed 19,000 people, consumed nearly two million tons of raw materials each year, and produced finished goods worth £2.5 million. Among the industry's by-products was a highly corrosive hydrogen chloride gas that, when exposed to moisture in the atmosphere, produced destructive clouds of acid. More than 13,000 tons of this "monster nuisance" were annually distributed throughout the English countryside, causing serious injury to crops and cattle. Several farmers filed suits against the alkali manufacturers. Some of these suits were successful, and while the damages awarded were generally not large, a few firms were forced to relocate. By the 1850s, however, it was becoming increasingly difficult to mount successful actions against the industry, since an increase in the number of producers made it virtually impossible to identify the precise firm that was responsible for the damage—a requirement under the common law.

In 1861 a prominent landowner, Sir Robert Gerard, angered by the steadily deteriorating value of his land, urged a politically influential neighbor, Lord Derby, to propose a legislative remedy. Derby, who at the time was prime minister, estimated the annual loss to his own estate and that of the other gentry in the area at £200,000. A select committee of the House of Lords was soon established, with Lord Derby as its chairman, to investigate the problem. Scientists testified that while a means for reducing emissions did exist, the inadequacy of the nuisance laws gave the manufacturers no incentive to adopt it. The manufacturers, many of whom had in fact already adopted a kind of crude condensing device, were not opposed to legislation making its installation compulsory provided that the costs of compliance were not unreasonable.

The Alkali Works Regulation Bill, introduced into the House of Lords on behalf of the government in April 1863, essentially followed the recommendations of the Derby Committee. While it did not specify any particular abatement technology, it did require that emissions of hydrochloric acid from each plant be reduced by at least 95 percent. The act was supported by the manufacturers, who regarded it as a way of protecting them from future common law actions, and it received the royal assent three months later. Its passage was significant in two respects. First, it represented the first time that Parliament had enacted regulatory legislation for the protection of property rights; all previous regulatory statutes had been directed at protecting people. Second, it marked the beginning of the central government's direct involvement in the control of pollution. To enforce this statute, Parliament decided to establish a separate inspectorate, to be administered by the central government itself.

Robert Angus Smith, considered the most celebrated sanitary chemist in the country, was appointed to head the Alkali Inspectorate. His four assistants likewise had extensive chemical training. Working closely with the manufacturers—indeed, Smith and his assistants regarded themselves essentially as consultants to industry—they showed them how to condense their gases more effectively and also devised a mechanism that enabled emissions to be continuously monitored. Within a year and a half the average escape of hydrogen chloride had been reduced to 1.28 percent, well within the legal requirement. By 1865 the emissions of acidic fumes had been reduced from 13,000 to 43 tons per week, and, as one pleased observer put it, "roses . . . grew where none had grown for years, and fruit trees had begun to blossom after having long ceased."[55] The chemical manufacturers were also pleased by the efforts of the Inspectorate: Smith had devised a means of recovering the waste produced by the manufacture of alkali that ultimately made the waste more commercially valuable than the original product itself. In 1868 Parliament, with the full support of the chemical industry, voted to continue the 1863 act indefinitely, and six years later the act was significantly strengthened with the support of both manufacturers and landowners.

What is noteworthy about this history is not simply Smith's achievement in both reducing alkali emissions and increasing industry profits but his strategy for doing so. Smith, like most of his counterparts in the United States a generation later working for both state and local governments, placed considerable emphasis on securing the cooperation of the industrialists whose conduct he was attempting to change. Eric Ashby and Mary Anderson write:

> It is not difficult to imagine the obstacles Smith had to overcome. . . . His only hope was to secure the confidence and cooperation of the factory owners. One tactless letter, one injudicious prosecution for infringement of the Alkali Act—and Smith would have had the whole alkali industry ganged up against him. . . . Yet to be lax and indulgent in the enforcement of the law was out of the question for a man like Smith.[56]

Smith subsequently recalled:

> There are two modes of inspection, one by a suspicious opponent, desirous of finding evil and ready to make the most of it. The other is that of a friendly advisor, who treats those whom he visits as gentlemen desirous of doing right. . . . The character of the inspection which I have instituted is one partially caused by my own inclination, and partly by the nature of circumstances.[57]

Smith, adopting the approach of the Factory Inspectorate, was extremely reluctant to prosecute violators of the law. Like his counterparts in American cities during the Progressive Era, Smith found that this policy created considerable controversy; he was regularly accused of being too friendly with the chemical manufacturers. He refused, however, to change his approach. Smith wrote in 1872: "Some of the public would have preferred to see [the Alkali Inspectorate] frequently in courts with cases of complaint, but I knew well that . . . habits must grow, and that to torment men into doing what required much time to learn was to return to the old system of teaching by the cane instead of by the intellect."[58] The Alkali Inspectorate, while clearly dealing with a unique problem, nonetheless had helped define the principles on which much future health and safety regulation in Britain would be based.

## THE ROOTS OF DIVERGENCE

Unfortunately, we lack a comprehensive history of the pattern of American regulation of corporate social conduct between the Progressive Era and the 1960s, in large measure because most regulation took place at the state and local levels.[59]

During the 1960s, however, the American approach to social regulation changed substantially in a number of critical policy areas. Some of this change took place at the local level: in the area of environmental protection, for example, some cities and states adopted a more aggressive enforcement strategy, attempting to implement rules and regulations that were strongly opposed by industry. The most important changes, however, were associated with the shift in regulation from the states and cities to the federal government. On the whole, federal regulation tended to be noticeably more aggressive in such policy areas as environmental protection, consumer protection, occupational health and safety, and equal employment than the state and local efforts had been.

Many of the most important differences between the British and American approaches to environmental protection in particular and social regulation in general are products of the last two decades. The active role played by the courts; the passage of detailed and highly specific statutes; the restrictions of the discretion of regulatory officials; the willingness of officials to prosecute; the introduction of strict timetables, specific goals, and technology-forcing standards; the considerable opportunities for participation by nonindustry constituencies; the lack of public trust in technical and scientific expertise—none of these phenomena was by any means unknown in the United States before the 1960s. Nor have they been utterly foreign to the British regulatory experience.

But on balance, they have become much more common in the United States since the mid-1960s.

Why, then, did the British approach to the making and enforcement of social regulation in general and environmental protection in particular remain relatively cooperative, while the American regulatory system became so much more adversarial? Why, when confronted with roughly similar problems and issues, did one country respond by maintaining the approach toward regulation that it had developed more than a century earlier—one based essentially on voluntary compliance—while the other transformed its strategy for controlling corporate social conduct and in doing so made government regulation as contentious as it had been throughout much of the nineteenth century?

The answer to these questions hinges on the responses to several others. Why was Great Britain able to develop and maintain a competent and highly respected civil service, while in twentieth-century America public officials have rarely enjoyed the confidence or trust of either business executives or reformers? Why did the British business community continue to regard government intervention in the economy as legitimate, while in America business attitudes toward government over the last half-century have remained relatively hostile? Why did the British middle class come to accept the notion that it was possible for business and government to work together in the public interest, while in America the consultation of industry by government officials became associated with the betrayal of the public interest? In short, what was there about the political and social context of the changes in regulatory administration instituted in the latter third of the nineteenth century in Great Britain that gave them a resiliency that has proved lacking in the United States?

The mid-Victorian reforms of regulatory administration proved viable because they rested on three components: a highly respected civil service, a business community that was prepared to cooperate with government officials, and a public that was not particularly mistrustful of large corporations. These three components are interrelated; each stems from the relatively subordinate role played by business and business values in British society and culture. The contemporary American style of social regulation is linked to the extent to which America—in sharp contrast to Great Britain—has remained very much a business civilization, a nation whose business community remains suspicious of public authority and whose public has little confidence in either the ability or willingness of government officials to control corporate conduct effectively.

A number of scholars, most recently Martin Weiner, have argued that Great Britain in the last part of the nineteenth century witnessed an important cultural backlash against the values of industrial civilization.[60] Just as England was nearing the pinnacle of its industrial strength, the British business community suffered a loss of self-confidence. Instead of seeking to expand its political and cultural influence, it became co-opted by a resurgent aristocracy; rather than continuing to emphasize industrial growth and international competitiveness, it now sought economic stability and social acceptability. Whether or not this decline in "industrial spirit" was a cause or a result of Great Britain's relative economic decline is less important than that the two developments accompanied one another. As Weiner notes, "at the very moment of its triumph, the entrepreneurial class turned its energies to reshaping itself in the image of the class it was supplanting."[61] Paralleling this decline in the prestige of business was a proportionate rise in the status of the professions, including that of civil servant.

These changes in social and cultural values were closely connected to the emergence of the mid-Victorian style of government regulation. The drive on the part of British industrialists for social acceptability helped weaken their resistance to expansion of public authority; accepting "reasonable" constraints on profit maximization was an important way of demonstrating that one had successfully transcended one's bourgeois origins and had become a gentleman. This was particularly true in the area of conservation, where, according to the "ethos of later-Victorian Oxbridge, . . . industry was noted chiefly as a despoiler of country beauty."[62] As British businessmen acquired country estates, they began to adopt some of the landed aristocracy's attitudes toward despoliation of the countryside by industry. More fundamentally, the relatively high social status enjoyed by civil servants in the latter third of the nineteenth century facilitated cooperation between the two sectors. It enabled regulatory officials to secure compliance by exerting social pressure.

The climate of public opinion during this period reinforced the development of a system of regulation based in large measure on informal social (as contrasted with strictly legal) controls, or, as one official put it, on the assumption that manufacturers are "gentlemen desirous of doing what is right."[63] The Victorian intelligentsia and professional middle classes may have been critical of the values of industrialists, but they were not troubled by either their economic or their political power; fears of monopoly or of the corrupting influence of industrialists on public policy are strikingly absent from the writings of Victorian social reformers. (Charges of corruption were more commonly leveled against the aristocracy.) While they recognized that the support of particular industrial interests was critical if their efforts to increase the scope of government controls over business were to succeed, they did not perceive the need to establish political mechanisms that either limited or checked the role of industry in shaping public policy. Rather, they appeared to be fully confident that once an abuse was exposed, professionally trained civil servants, acting within the broad framework of legislation enacted by Parliament, would be able to formulate a remedy and persuade the business community to adopt it. In short, they saw their roles as those of educators rather than policemen.

The British effort to develop a modern administrative state during the latter third of the nineteenth century was also facilitated by the fact that Great Britain's traditions of public authority were relatively strong to begin with. As Daniel Bell has suggested, Great Britain became a capitalist society without ever becoming a bourgeois one.[64] Notwithstanding the ideology of Manchester liberalism, Great Britain never came close to approximating the ideal of the "night watchman" state; indeed, Spencer's ideas were far more influential in the United States than in Great Britain. Particularly at the local level, the landed gentry continued to exercise important governmental functions throughout the eighteenth and nineteenth centuries. Indeed, as Joseph Schumpeter writes, "the aristocratic element continued to rule the roost right *to the end of the period of intact and vital capitalism.*"[65] Thus British public administrators enjoyed an important element of continuity and were able to draw on a tradition of deference toward public authority on the part of both industry and the public.

The political climate of Progressivism differed in important respects. The American civil service movement did take its model "for reform and the new American state from Great Britain, where the civil service had been successfully restructured and where policy-oriented parties, commercial classes and Oxbridge intellectuals had recently consolidated their hold over the central government."[66] However, the American effort to build a public bureaucracy staffed with highly qualified personnel and provided with sufficient authority to supervise the activities of industry met with major obstacles.

Among the most important was the attitude of business. The views of the American business community toward public authority did undergo an important change during the Progressive Era; a faith in social Darwinism was replaced by a more sophisticated understanding of the appropriate role of government in a modern industrial society. Yet despite their increased willingness to acknowledge their "social responsibilities," the underlying values of the American industrial elite were not transformed; they acquired, in Robert Reich's phrase, only a "veneer of civic virtue."[67] In sharp contrast to their counterparts in Great Britain, they remained firmly committed to profit maximization, industrial growth, and economic expansion. There was no aristocracy to co-opt them, and the impact of the social gospel movement on the moral and social outlook of American businessmen pales by comparison with that of Victorian evangelism on industrialists in Great Britain. Consequently, the ideology of "corporate liberalism," with its acceptance of a legitimate role for public authority and its notion of a partnership between "big business" and "big government," never really took root in American business culture; American executives remained both extremely jealous of their autonomy and highly mistrustful of public authority.[68]

Not surprisingly, public service never managed to acquire the social status that it came to enjoy in Great Britain. As a result of the introduction of the civil service system, the quality of officials in the public sector improved markedly around the turn of the century. After having virtually absented themselves from public life during the Gilded Age, members of the nation's social elite—Theodore Roosevelt is the most prominent example—began to seek careers in the public sector. But among Americans, in contrast to the British, business did not suffer a proportionate loss of prestige and moneymaking lost little of its social appeal. While in Britain the professions and the civil service offered an attractive alternative to employment in industry—and thus were capable of attracting first-rate talent—in America the pursuit of wealth through economic activity remained the dominant interest of the nation's upper class. As a result, civil servants in America never acquired the status that they did in Great Britain. Far from viewing them as social equals, American business executives continued to regard them as their social and intellectual inferiors. They might cooperate with them—when such cooperation was in their self-interest—but they hardly felt under any social pressure to defer to either their authority or their expertise. In short, notwithstanding the efforts of the Progressives to "civilize" the nation's business community, America remained very much a business civilization.

At the same time, the attitude of many Progressives toward the large business corporation differed significantly from that of their counterparts across the Atlantic. Precisely because American society and politics had been so dominated by the large corporation during the latter half of the nineteenth century, public opinion during the Progressive Era became highly critical of "big business." The Victorian intelligentsia and much of the middle class may have looked with skepticism on the purposes of business civilization, but they were not antibusiness in the sense that they mistrusted its power and feared its domination of the political process. For reformers in America, the opposite was the case. They had no quarrel with the fundamental objectives of industrial society; it was precisely the power of business within that society that troubled them.

Moreover, the American middle class had no aristocracy to help it limit the political influence of industry. While the degree of industrial concentration was substantially greater in Great Britain, only in America did the trusts become a symbol of the corruption of the commonwealth by wealth. Louis Hartz writes:

> Granted that America now superseded England as the home of the "great industry,". . . it is still a fact that the relative concentration of economic power was greater in almost any part of Europe than it was in America. And yet the European Liberal reformers, though they blasted "monopoly"—the English in the case of the tariff and land, the French in the case of large business in general—did not make the same fetish of the symbol that the American Progressives did.[69]

If socialism, as Hartz argues, emerged as a reaction to the vestiges of feudalism in capitalist society, then populism represents the ideology of opposition to the economic elites of a capitalist society in which those remnants are virtually absent. The energies that in Great Britain and other European nations went into challenging the power of the landed aristocracy, the clergy, and other pillars of the establishment—or even more basically, into efforts to extend the franchise—in America were directed disproportionately at the prerogatives of the business corporation. Precisely because America represented virtually a textbook case of capitalist development, in which the first large national institution to emerge was the business corporation, the corporation in America became the primary focus of populist pressures.[70] In short, the development of industrial capitalism in America stripped away all the protective veils around the corporation; it alone came to symbolize the injustices of the industrial/capitalist order.

As a result, while individual businessmen may have been strong supporters of many reforms enacted during the Progressive Era, the ideology of much of the movement was extremely hostile to large companies. Given their perception of the political strength of business—and of the magnitude of the gap between the power and status of business and government—many Progressives were unwilling to trust government officials to supervise the activities of the private sector effectively. Accordingly, they sought to reduce the power of the "interests" and strengthen the power of the "people" through such devices as reducing the size of corporations, increasing public disclosure, limiting corporate funding of federal election campaigns, and establishing the initiative and the referendum at the state level.

Moreover, the effort to develop a "modern administrative state" in the United States at the beginning of the twentieth century suffered from another important handicap: compared with Great Britain's, both public authority and the administrative strength of the public sector were far weaker to begin with. If Great Britain experienced a capitalist revolution without ever having become a bourgeois society, then America was a business civilization even before it was a capitalist one. In the first few decades of the nineteenth century, the tradition of public service exemplified by the Founding Fathers fell into disrepair; it was replaced by a spoils system in which public service became an alternative means of private gain. As Stephen Skowronek demonstrates in *Building a New American State*, throughout the nineteenth century America was governed by a combination of parties and courts—neither of which provided any basis for a coherent system of public administration. In addition, a mistrust of public authority in general and of professional values and expertise in particular has deep roots in American political culture, even predating industrial capitalism. Such attitudes, as described by Samuel Huntington, clearly represented a substantial obstacle to the construction of a system of regulation based on administrative discretion and legitimized by professional expertise and administrative neutrality.[71]

Consequently, civil servants in the United States, including officials of the newly established regulatory agencies, were never able to acquire either the power or the prestige of their counterparts across the Atlantic. While courts in Great Britain imposed relatively few restraints on government administration, judicial activism in the United States—initiated primarily by business as a way of limiting the scope of government regulation—increased during the late nineteenth and early twentieth centuries. While in Great Britain the heads of important regulatory agencies such as the Alkali Inspectorate were civil servants, policy-making positions throughout America's regulatory bureaucracy continued to be filled by political appointees. And while Parliament came increasingly to delegate its authority to ministers—thus the civil service—both the American Congress and the state legislatures remained relatively active in the making of regulatory policy, though their role was significantly smaller than it has become in recent years.

## THE ROLE OF BUSINESS

The fragility of the Progressive mode of regulation—as contrasted with the more enduring legacy of mid-Victorian reform efforts—became evident in subsequent decades. In critical respects business–government relations have been relatively cooperative in Great Britain over the last century. The late-nineteenth-century shift in cultural values that reduced the prestige of industrialists and enhanced that of civil servants has proved permanent. Within the business community itself the process of socialization has continued. "More and more in the twentieth century, the higher echelons of the larger businesses were dominated by men whose standards had been formed in the gentlemanly mode."[72] And to the extent that Great Britain has a populist tradition, it has continued to focus on preindustrial symbols of authority, such as the House of Lords and the other trappings of aristocracy, rather than on industry. Great Britain certainly has had a politically vigorous and in some respects radical trade union movement, but with a handful of exceptions, it has not focused on regulatory issues.

The contrast with American history since the Progressive Era is marked. During the 1920s business was able to reassert both its power and its prestige. Business–government relations during the 1920s may have appeared to be based on cooperation, but it was clearly not cooperation among equals: business's acceptance of public authority was no greater then than it had been during the last third of the nineteenth century. The ideology of corporate liberalism atrophied during the "dollar decade," and Herbert Hoover's idea of an "associative commonwealth" never attracted significant support.[73] Moreover, the commitment of the nation's socioeconomic elite to public service also proved ephemeral: following World War I they returned to the private sector. When a modern administrative state was finally established in the United States during the New Deal, it met with fierce opposition from those who perceived it—correctly—as threatening their power and privileges. In sharp contrast to the situation in Great Britain, business–government relations during the 1930s in the United States were highly adversarial; the increase in government controls over business that took place during the New Deal was associated with a reduction in business's political influence and autonomy in important policy areas.

The relative willingness of the British community to cooperate with regulatory authorities in a variety of areas has been symptomatic of a much broader acceptance of public authority. Over the last century, British business elites have proved far more willing to accept a whole range of social and political constraints than have their counterparts in the United States. British industrialists accepted the legitimacy of trade unions far more readily than did business executives in the United States: American industrialists were engaging in violence against workers who were attempting to join unions more than forty years after they had become an established part of the British system of industrial relations. The welfare state measures introduced by the New Deal were bitterly opposed by the overwhelming majority of corporate managers, while the far more extensive expansion of the welfare state under the Atlee government immediately following World War II was adopted on the basis of widespread consensus.[74] Similarly, the extremely limited efforts of the New Deal to expand government ownership were vehemently resisted by business executives in the United States, while the far more extensive nationalizations undertaken by the postwar Labour government—with the exception of iron and steel—met with remarkably little business opposition.[75]

My argument is not that business–government relations in the United States have been consistently adversarial while the relations between the two sectors in Great Britain have been uniformly cooperative. Certainly, economic regulation in the United States has been characterized by a high degree of cooperation between regulators and the industries they were responsible for regulating; the relationship between the Defense Department and defense contractors has likewise been mutually supportive over the last forty years. Even during the 1970s, the relationship between social regulatory agencies and industry was not uniformly adversarial. In Great Britain, on the other hand, many aspects of government intervention in the economy, particularly since World War II, have been strongly criticized by industrialists, and the British business community's high appraisal of the good judgment and technical competence of regulatory officials certainly does not extend to those civil servants responsible for promoting the competitiveness of the British economy.[76] But for all its grievances, the British business community exhibits a degree of deference toward the norms and values of the civil service that has no equivalent in the United States. As J. P. Nettl writes, "lacking a firm sense of their distinct identity, and belief in their distinct purpose, businessmen have been particularly vulnerable to the pressure of the consensus as emanating from Whitehall."[77] Edward Mason adds: "A British businessman can say 'Some of my best friends are civil servants' and really mean it. This would be rare in the United States."[78]

The economic and political values of each nation's business community are related. Business executives in America continue to regard government officials much as they view their competitors: as challengers to be met as aggressively as possible. British businessmen appear less competitive with each other and more willing to cooperate with government. To the extent that the only standard of success for American businessmen remains the "bottom line," they have tended to regard any restraint on management's prerogatives as threatening.

While the highly individualistic ethos of American business culture has severely limited the role of trade associations as a vehicle for industry self-regulation, in Great Britain trade associations represent an important source of peer pressure on firms to behave "responsibly." The term the British use to describe a company that deviates from industry norms is "cowboy"—one clearly drawn from American popular culture. And while trade associations have assisted the British government in enforcing various regulatory policies—transmitting the views of the government to industry is in fact an important purpose of British trade associations—such organizations have emerged as an important focus of opposition to government regulatory initiatives in the United States.

Ironically, while the ideology of corporate social responsibility is far more widespread in the United States than it is in Great Britain, the practices of British business actually appear to conform much more closely to its ideals.[79] This is not to suggest that the British business community is any more virtuous or has not on occasion treated its consumers, workers, or the environment every bit as irresponsibly as its counterpart in the United States. Rather, it is to suggest that the norms under which it operates make it more susceptible to social pressure from both government officials and other firms to behave "responsibly." Alternatively, the relative formalism of the American regulatory system—its reliance on clearly defined rules and standards—reflects the inadequacy of informal mechanisms of social control within a highly individualistic culture: the *Federal Register* represents America's substitute for the Queen's Honours List.

## The Contemporary Setting

The contemporary changes in the pattern of government regulation in the United States cannot be understood apart from the substantial shift in public attitudes toward business that occurred during the second half of the 1960s—a shift rooted in the dynamics of the relationship among business, government, and the public that has just been described. After more than two decades during which the American public was relatively supportive of the purposes and prerogatives of the large business corporation, public attitudes toward business became relatively hostile. Paralleling this change in public opinion was a substantial increase in the scope of government regulation over a wide variety of aspects of corporate social conduct: between the mid-1960s and the mid-1970s— the period when the major initiatives in environmental policy took place— more regulatory legislation was enacted and more new regulatory agencies were established to administer them than in the entire previous history of the American federal government.

These laws and agencies were by no means unanimously opposed by the American business community, and individual companies and industries have certainly both influenced and benefited from the way particular regulations have been written and enforced. Nonetheless, these sweeping changes in regulatory policy were associated with a significant reduction in the political influence of business in the United States and a corresponding increase in the power of public interest organizations that were relatively critical of business.[80] The result was what Murray Weidenbaum has labeled a "second managerial revolution," a significant transfer of authority over a number of critical corporate decisions from professional managers to government officials.[81]

The increase in public interest in environmental regulation during the late 1960s was a global phenomenon. What was distinctive about its American variant was the way in which the issue of pollution control came to be defined. Essentially the debate over environmental regulation represented a contemporary version of American populism: the interests of "big business" in production and pollution were contrasted with those of the "people" in the preservation of the ecosphere. To exaggerate only slightly, in America during the late 1960s and early 1970s, pollution was blamed on business. One journalist, writing in 1968, described the public's perception of the cause of pollution as dominated by the "economic giants of the steel industry, the power industry, the petroleum industry, the chemical industry, the pulp and paper industry, and many lesser enterprises."[82] The data from polls taken between the mid-1960s and the early 1970s reveal that the public both held industry responsible for pollution and believed that its efforts to clean up the nation's air and water were inadequate.[83] Anthony Downs wrote in 1972 that "much of the blame for pollution can be plausibly attributed to a small group of villains whose wealth and power make them excellent scapegoats."[84] Not only was the pursuit of profit regarded as responsible for the deterioration of the environment, but business was assumed to possess sufficient financial and technical resources to be able to improve the nation's physical environment as rapidly as the public demanded.

Accordingly, the political influence of industry was invoked to explain the ineffectiveness of the previous regulatory efforts at both the local and national levels. For environmentalists and many of their supporters, the history of government regulation was a tale of failure. Thus, if future environmental regulation were to be effective, what was required was not simply an expansion of the scope of government controls but rather a series of critical changes in the way regulatory policy was made and enforced—changes that reduced the influence of business and increased that of political constituencies that were committed to the "public interest."[85]

This perspective helps make sense of many of the distinctive features of American environmental regulation and, by extension, social regulation in general. Consider, for example, the highly controversial decision to focus a disproportionate amount of the nation's pollution-control efforts on the reduction of automobile emissions; in no other nation has pollution from mobile sources received so much attention as in the United States. Moreover, virtually the entire burden for reducing automobile emissions has been placed on Detroit; the federal government has been either unwilling or unable to require local governments to place any restrictions on automobile use. Indeed, throughout the 1970s the EPA did not even make a serious effort to require that automobile owners adequately maintain—let alone not tamper with—the pollution-control equipment installed at the factory. Only the automobile manufacturer is held legally responsible for complying with automobile emission standards; federal legislation contains no penalties for automobile owners. It was not until 1984 that the EPA finally managed to force California— the very state whose concern with automobile pollution prompted federal regulation in the first place—to require automobile owners to have their pollution-control equipment checked annually.

Neither the decision to control automobile emissions more strictly than pollution from any other source (motor vehicles were the only pollution source in which specific numerical reductions were actually mandated in the 1970 Clean Air Act) nor the way in which pollutants from automobiles have actually been restricted can be explained on environmental grounds. The Clean Air Act Amendments of 1970 were enacted in the absence of any scientific evidence as to the dangers to public health posed by automobile emissions, and subsequent evidence suggests that they have been among the most cost-ineffective environmental regulations adopted by the federal government.[86] For most Americans, emissions from stationary sources, particularly utilities, represent a far more substantial threat to both health and environmental quality. The priority given to Detroit by the federal government in 1970 reflected widespread public resentment against the automobile industry—an industry that, by virtue of its size, concentration, and profitability, had come to epitomize American big business: it was held largely to blame for the deterioration of the nation's air quality and thus made uniquely responsible for designing and installing the pollution-control equipment regarded as necessary to give Americans the air quality they had a "right" to expect.

This analysis also makes sense of the way in which the health issues associated with environmental regulation in the United States have been defined. In no other industrial nation has the problem of pollution control been so closely identified with disease as in the United States. The United States is the only industrial nation to have a separate set of regulatory standards for carcinogens and the only nation to attempt to develop a zero-risk threshold for various pollutants and occupational hazards. Moreover, the notion that "industry causes cancer," though not supported by any scientific evidence, has enjoyed wider currency in the United States than in any other country.[87] Indeed, over the last decade the American public has become preoccupied with the threats posed by industrial activity to its health and safety to a far greater extent than the citizens of any other industrialized nation. Mary Douglas and Aaron Wildavsky write:

> Try to read a newspaper or news magazine, listen to radio, or watch television; on any day some alarm bells will be ringing. What are Americans afraid of? Nothing much, really, except the food they eat, the water they drink, the air they breathe, the land they live on, and the energy they use. In the amazingly short space of fifteen to twenty years, confidence about the physical world has turned into doubt. Once the source of safety, science and technology have become the source of risk. What could have happened in so short a time to bring forth so severe a reaction? How can we explain the sudden, widespread, across-the-board concern about environmental pollution and personal contamination that has arisen in the Western world in general and *with particular force in the United States?*[88]

This phenomenon cannot be explained by the magnitude of the actual physical hazards that confront Americans; by international standards American cancer rates are relatively low. Nor are they increasing at a faster rate than those of other nations. Not only do American longevity statistics compare favorably with those of other industrial countries, but during the 1970s—the very period when public anxiety over the health hazards of technology reached its apogee—the average American's life expectancy increased more than in any comparable time period.[89]

Why, then, have regulatory officials been under pressure to adopt—and have frequently adopted—regulations reflecting a more conservative assessments of risk in America than in any other industrial society, including Britain? My explanation (though not Wildavsky's) is that the contemporary American public's suspicion of technology is directly linked to both its hostility to business and its mistrust of government: it is not so much science and technology per se that the American public finds threatening as the extent to which they appear to be controlled by private companies—companies that are regarded as both inherently irresponsible and too powerful to be effectively regulated by government. Threats to the public's health and safety have not been seen, as they are in Britain, as an inevitable component of production and consumption in a highly industrialized and affluent society; rather they have become identified with the profit motive of America's largest firms.

The controversy in America over the use of cost-benefit analysis in the making of regulatory policy can be understood in similar terms. Only in the United States has cost-benefit analysis emerged as a major political and legal issue. In every other country, including Great Britain, policy makers are aware that it is impossible to quantify the benefits of regulatory controls; by definition, social regulation involves a choice among competing values. Yet they are equally aware that government regulation is not costless and that therefore in a world in which resources are limited, it makes sense—at least at the margin—to attempt to make some assessment of the costs and benefits of particular rules and restrictions. Why, then, has the effort to use cost-benefit analysis as a guide to the making of regulatory policy created so much controversy in the United States? Cost-benefit analysis has been strongly opposed by many pro-regulation constituencies in the United States because they view it as a way for industry to justify its unwillingness to make expenditures they believe it should make. They dislike it because it serves to legitimate the interests and values of business in the making of regulatory policy. Much the same could be said of the science of risk assessment, which was originated in Great Britain and has been employed extensively there, but whose introduction in the United States has been repeatedly challenged.[90]

A similar analysis can be made of two other unique features of American environmental regulation: the use of statutory deadlines and technology-forcing standards. No other nation has attempted to establish such strict timetables for the attainment of various environmental objectives or to enact emission requirements before the technology has been invented to comply with them; nor has any other nation legislated environmental objectives whose achievement is—for all practical purposes—impossible (although the European Community is now moving in all three directions). Each of these decisions was based on the assumption that the main obstacle to adequate compliance was political rather than economic or technological; if regulatory requirements were made sufficiently strict, then industry somehow would manage to find the resources to provide the public with the environmental protection it deserved. By including these goals and standards in legislation, environmentalists and their supporters hoped to make it more difficult for industry to use its political influence over regulatory agencies to delay or prevent their implementation.

The American approach to social regulation has not rested simply on holding industry disproportionately—if not exclusively—responsible for the nation's environmental and health and safety problems; it has also attempted to keep regulatory officials and industry representatives from collaborating on their solution. The transformation of the American approach to regulation in the early 1970s was motivated largely by an effort to reduce industry's ability to participate in the making and enforcement of regulatory policy. For environmentalists and other pro-regulation constituencies, there was no middle ground between capture and coercion. The notion of cooperation between business and government was inherently suspect; the less business executives were able to interact with regulatory officials, the more likely regulations that were actually in the public interest could be devised and enforced. Given many Americans' perceptions of the disproportionate power exercised by large companies and the strength of the business ethic within American culture, capture was seen as inevitable in the absence of strong safeguards. Eugene Bardach and Robert Kagen write:

> In the most significant regulatory areas, the law has been deliberately structured to prevent capture, to program inspectors to apply regulations strictly, to pressure enforcement officials to apply formal penalties to violators and to adopt a more legalistic and deterrence-oriented stance vis-à-vis regulated enterprise.[91]

In this sense, conflict between industry and government was deliberately designed into the American system of governmental regulation in the late 1960s and early 1970s. For many supporters of regulation, the very magnitude of industry opposition was the most convincing evidence that their approach to regulation was working. Too much cooperation by industry would have suggested that enforcement was too lax. While in Britain a resort to prosecution is viewed as reflecting a failure on the part of enforcement officials, in America it became an important index of their integrity. In short, much of what made contemporary social regulation seem unreasonable in the eyes of the business community was precisely what gave it credibility in the view of activists and much of the public.

In the area of pollution control, for example, the establishment of technology-forcing and uniform emission standards, strict deadlines for compliance, and provisions for judicial review of agency decisions all represented an attempt on the part of Congress to make the EPA's relationship with industry as "arm's length" as possible. The EPA officials appointed by the Nixon administration were certainly not particularly antagonistic to business. Indeed, in 1969 the administration established the National Industrial Pollution Control Council, consisting of representatives of major companies and trade associations, precisely in order to "provide a means for American industry to work cooperatively with the government in pollution control and abatement."[92]

But regulatory officials in America immediately found themselves under intense pressure from both Congress and members of the public to demonstrate their "independence" from the business community; much of the EPA's initial efforts at strict enforcement were undertaken to maintain its credibility with the political constituencies that had led to its creation in the first place. And when these pressures proved inadequate, environmentalists filed a steady stream of lawsuits aimed at diminishing the EPA's administrative flexibility, thus further driving a wedge between agency officials and business. The distinctively cautious attitude of American regulatory authorities toward risk can be understood largely as a response to those same legal and political pressures. In the political climate of the 1970s, regulatory officials found themselves far more vulnerable to public criticism on the grounds that their controls on industry were insufficiently stringent than that they were too severe.

The political and ideological context in which British environmental policy was implemented during the 1960s and 1970s was substantially different. British industry may well have found itself under severe political and economic pressures during much of the 1970s, but these pressures had little to do with government regulation. The increased political saliency of environmental protection was not associated with an increase in public hostility toward business, as it was in America. Christopher Wood writes:

> While industrial pollution problems exist, of course, there is no general belief that industry is the sole cause of pollution. Many people remember how bad smog (smoke plus fog) was and remember changing from coal fires to cleaner fuels. They participated in the improvement they witnessed. In America, on the other hand, though widespread changes in heating patterns have taken place . . . the general perception is that pollution is industry's affair, not the people's. . . . This difference in view is crucial—the British appear to accept that the consumption of goods, as well as their production, causes air and other types of pollution: the Americans seem very reluctant to recognize this.[93]

The British public is less likely to regard regulation as a zero-sum game, in which every policy or decision in favor of industry invariably means the public has "lost" and vice versa. Nor do the British view their history of government regulation as a tale of failure; on the contrary, they believe that considerable progress has been made in safeguarding and improving the quality of the environment over the last quarter-century. Moreover, to the extent that this progress has been disappointing, they do not ascribe it primarily to the lack of integrity of regulatory officials or the unwillingness of industry to devote adequate resources to pollution control; rather, the British public appears to be cognizant of the extent to which the effective enforcement of environmental regulations has been hampered by the lack of both private- and public-sector funds.

The extensive consultation of industry by government officials before regulatory policies are made and enforced has rarely aroused controversy or criticism; in Great Britain such interactions generally appear to be regarded as appropriate. Rather than seeking to limit "conflicts of interest," the British system of public administration is instead based on them: both government officials and industrialists frequently wear multiple hats. Indeed, a civil servant who did not have extensive ties with executives in the industries he was responsible for regulating would be regarded as derelict in his duties. For the most part, even critics of British government regulation have not insisted that industry and government be kept at arm's length. While some may resent industry's privileged position in the regulatory process, they do not automatically equate cooperation with capture. The British public, to a far greater extent than the American, perceives governmental officials as having sufficient integrity and competence to be able to represent its interests reasonably well, even if not all governmental decisions are subject to public scrutiny.

I earlier suggested that confrontational and cooperative approaches to regulating industry may be equally effective. This conclusion, however, must now be qualified. They may be equally effective across national boundaries, but they are not necessarily equally effective within any given society. And in fact they are not.

The decision of American environmentalists and many of their supporters to give high priority to challenging the political influence of industry over the regulatory process was a sound one. Americans have good reason to be suspicious of the power of business and the motivations of business executives. Given their perception that the history of government regulation of business in America was a history of industry dominance—an analysis documented repeatedly by studies of business–government relations published during the 1960s—the strategy adopted in the 1970s was not an unreasonable one. The capture theory of government regulation of business may be misleading as a description of

257

business–government relations during the 1970s, but it is certainly not inaccurate as a historical account of much government regulation in the United States.

Had the pattern of interaction between business and government remained cooperative, the United States might well have made less progress in improving the quality of its physical environment over the last fifteen years. For all its myriad shortcomings, the adversarial relationship has worked; it has undoubtedly forced industry to allocate far more resources to improving the quality of the environment than it would otherwise have done. Compared to many of the other domestic policy initiatives undertaken by the United States over the last two decades, environmental regulation must be judged as among the more successful. While it is undoubtedly true that similar—if not in some cases better— outcomes could have been achieved by a more flexible approach, it can also plausibly be argued that a less adversarial style of enforcement might well had led to less progress. Whereas in Britain the government's commitment to environmental protection is greater than the political influence of its environmental movement would suggest, the relative effectiveness of American environmental regulation does appear to be attributable in large measure to the ability of environmental pressure groups to countervail the political influence of industry.

The much-criticized use of technology-forcing standards, for instance, did significantly accelerate the development of abatement technology in numerous industries. One study concludes: "The Clean Air Act Amendment of 1977 induced substantial private emission-control innovations. Examination of the copper smelting and electric power industries—two industries pressed hard by the Act's requirements—proves that statutes can force technological innovation."[94] Congress's insistence on requiring industry to meet unrealistic deadlines has undoubtedly kept the public's attention focused on the gap between public intentions and private compliance.[95] It is not coincidental that the industry that has made the most progress in reducing its emissions is also the one toward which a disproportionate amount of the government's enforcement efforts have been directed: the automobile industry. Certainly the behavior of this industry during the 1950s and 1960s offered little ground for public confidence in its willingness to devote adequate resources to designing and installing pollution-control equipment in the absence of intense political and legal pressures. Shortly before the debate over the 1970 Clean Air Act Amendments began, the industry had signed a consent decree with the Department of Justice that "at least implied that the big three manufacturers had in fact illegally worked together to thwart air pollution control."[96] As one senior Senate staff committee member stated in 1970, "the industry's statement before this committee as to what they are capable of doing, and their performance in California in claiming that the state standards could not be met, have made us skeptical of what they say."[97] And can anyone doubt the magnitude of the enforcement efforts that would be required to reduce sulfur emissions from midwestern utilities?

In a sense, each nation's business community has experienced the kind of regulation it deserves. The British approach is viable precisely because British business does not have a confrontational attitude toward public authority. It can be trusted to negotiate in good faith. The American approach toward enforcement assumes bad faith on the part of both parties: it is uniquely suited to a business community that is extremely competitive, jealous of its prerogatives, and contemptuous of government officials. And government officials in both societies get the response of industry that they deserve. British industrialists are relatively cooperative because the demands imposed on them are reasonable; American executives are frequently antagonistic to government officials precisely because many of the demands imposed on them are not.

## FUTURE CONVERGENCE?

Are the regulatory policies of Great Britain and the United States likely to converge once again? Some evidence certainly suggests that they may be doing so. The British government is finding itself under increasing pressure to reduce the secrecy that surrounds its regulatory policies—particularly those that directly affect public health and safety—and to make more explicit the criteria its officials use to determine whether or not firms are in compliance and whether or not a particular process, product, or emission constitutes a hazard. Moreover, the controversy surrounding the fairness of the large public inquiry reveals a lack of public confidence in what has become a critical dimension of British environmental decision making—though these inquiries rarely involve private industry.

Most important, Britain is finding itself under continued pressure to make its system of regulation conform more closely to those of the other member states of the European Community. As we have seen, Britain has already made changes in response to Community directives and will undoubtedly make additional ones in future years. Such changes will have the effect of making British and American regulatory policies more similar, since the European Community's approach to regulation resembles that of the United States both in its use of legislation to force the pace of change and in its reliance on precise timetables and centrally specified standards. (Both the EEC and the United States are, in effect, federal systems, and neither is in a position to rely on informal measures of social control to bring about desired changes in corporate conduct.)

At the same time, during the latter part of the 1970s the Carter administration initiated reforms designed to make environmental regulation in the United States more flexible. The two most important included the "bubble concept" and the trading of emission rights. The former, established in 1979, permitted the EPA to consider all the pollution coming from a particular plant as if it originated from a single stack, thus allowing each company to determine the most cost-effective strategy for reducing its emissions. The policy was subsequently expanded to allow for creation of multiplant and multi-industry "bubbles." The trading of emission rights allows for additional construction in nonattainment areas; if a firm wishes to relocate or expand its facilities in such an area, it can do so by paying an existing pollution source to reduce its emissions. A growing number of efforts have also been made to mediate environmental disputes—indeed, a private nonprofit organization has been established to provide such a service—in order to reduce the costs and delay associated with litigation. Finally, both the Carter and Reagan administrations have required the EPA and other regulatory agencies to consider the costs of compliance before promulgating new regulations.

While it would be wrong to dismiss either set of developments, their significance needs to be put in perspective. The British approach to environmental regulation may well become less distinctive with the passage of time, but we have little reason to expect regulatory policies and procedures so deeply rooted in a nation's political traditions and institutions to change substantially: the pace of change is likely to remain modest. Indeed, the very way in which British regulatory policy is made and implemented virtually ensures against any radical policy departures. Moreover, for all the controversy that continues to surround particular policies, government regulation of industry in general and of environmental issues in particular occupies a relatively peripheral place on Britain's political agenda; altering its approach to regulation remains among the least pressing of Britain's domestic concerns. Certainly, in comparison with the Reagan administration, the Thatcher government has made relatively little effort to initiate major changes in this policy area.

My description of American environmental policies has drawn heavily on the 1970s. Does the history of environmental regulation under the Reagan administration suggest that America has moved any closer to the British approach? Soon after the 1980 elections, EPA Administrator Ann Gorsuch and Assistant Administrator Rita Lavelle initiated a major change in the agency's enforcement strategy. Under their leadership, the agency deemphasized the filing of lawsuits and the issuing of enforcement orders; instead it sought to negotiate "voluntary" agreements with industry. This nonconfrontational approach "grew out of the administration's sympathy for the real challenges faced by . . . industry, and Gorsuch and Lavelle's apparent belief that industry was willing to act in the public interest."[98] It also reflected their view that a policy of cooperation was more likely to result in improved compliance, since it would avoid the delays occasioned by litigation. Defending the EPA's policy of raising funds from industry for the cleanup of toxic wastes through negotiated settlements, Gorsuch informed Congress:

> The purpose of prelitigation negotiation is to encourage cooperative responsible parties to clean up sites. Cleanup resulting from negotiation and settlement is likely to occur more quickly than cleanup resulting from litigation. . . . Negotiation can be particularly effective before litigation begins. The threat of litigation or administrative orders during prelitigation negotiations gives the federal government valuable leverage, because settlement would allow private parties to avoid the great expense and adverse publicity involved in litigation. This leverage is sacrificed if the federal government automatically files cases before it begins negotiations with responsible parties. In addition, filing lawsuits before negotiations even begin alienates potentially cooperative responsible parties.[99]

Whether or not the substitution of the policy of "shovels first, lawyers later" for one of "lunch now, lawyers maybe, but shovels never" did in fact accelerate the cleanup of toxic wastes remains unclear: negotiated settlements that some Superfund officials considered important accomplishments were perceived by others as sweetheart deals.[100] In any event, the effectiveness of the administration's new enforcement strategy rapidly became overshadowed by questions concerning its propriety. In late 1982 and early 1983 Congress became increasingly impatient with the pace of the EPA's cleanup effort. After Gorsuch's refusal to provide documents requested by a congressional committee, she, along with nearly all of the EPA's other political appointees, was forced to resign.

The history of environmental regulation under the Reagan administration demonstrates that regulatory policy remains highly politicized in the United States. Notwithstanding the more conservative political climate of the late 1970s and early 1980s, environmental policy has remained highly visible, highly controversial, and, perhaps most significant, highly unstable. If anything, the widening of the enforcement gap under the Reagan administration has expanded rather than reduced the unpredictability of public policy in this area. Moreover, the controversy created by the Reagan administration's initiatives in the area of environmental policy at both the EPA and the Department of the Interior suggests the political obstacles that stand in the way of making regulation more cooperative. While the failure of Gorsuch's reforms was due in part to her political ineptness, it is worth noting the kinds of interactions with industry for which she was so strongly criticized. Consider the following evidence of the EPA's politicization under Gorsuch, as revealed during an investigation by Congress.

> Decisions concerning when to award grants from Superfund to clean up hazardous waste sites and what techniques to use were often based upon the decisions' calculated effects upon state elections or upon business giving financial support to the president.
>
> High officials in EPA permitted the Dow Chemical Company to review a draft agency report on Dow's responsibility for dioxin contamination of surface waters near Midland, Michigan. The company also was permitted to suggest alterations in the report's language.
>
> EPA employees permitted representatives of the formaldehyde industry to review and comment upon technical studies concerning the hazardousness of the chemical in private meetings from which other interests were excluded.[101]

Certainly, the first practice would also be regarded as inappropriate in the United Kingdom. The latter two incidents, however, would be regarded in Britain as unexceptionable: they would be taken as evidence not of corruption but of consultation. Indeed, were Gorsuch a British civil servant, failure to consult with industry before issuing a study or regulation would have made her subject to criticism from Parliament.

The continued public suspicion of cooperation between regulators and the regulated in the area of public health and safety in the United States is also suggested by the experience of OSHA under the Reagan administration. Like Gorsuch, OSHA Administrator Thorne Auchter reduced his agency's emphasis on prosecution as a compliance strategy. Explaining that "our job is health and safety . . . we're not interested in crime and punishment,"[102] OSHA curtailed both the frequency of its inspections and the number of violations for which firms were cited: the dollar amount of the penalties imposed by the agency declined 69 percent between 1980 and 1982. At the same time, OSHA expanded its efforts to promote voluntary compliance: the annual budget of its on-site consultation program increased by more than 25 percent between 1980 and 1982. While occupational injury rates were stable during OSHA's first decade, between 1981 and 1983 they declined somewhat. The extent to which this reduction can be attributed to the shift in the agency's enforcement strategy remains unclear, since at least part of this decline is presumably due to the 1982–83 recession. In any event, Auchter came under strong criticism for substituting a more cooperative approach to compliance for a more coercive one; his critics argued that cutbacks on OSHA's enforcement efforts invariably made the workplace less safe.[103]

Thus under the Reagan administration, business and government may have become more cooperative, but the United States has certainly not moved any closer to the British model. For as the relationship between business and government became less adversarial during the first half of the 1980s, the relationship between government and the public grew significantly more so. The American public has not perceived the Reagan administration's changes in regulatory policy as making the process either more effective or more efficient. On the contrary, the officials appointed by the administration have been widely perceived as altogether too sympathetic to business. In this sense, the politics of American social regulation remains a zero-sum game: business's satisfaction with the administration's regulatory policies has been precisely counterbalanced by the public's unhappiness with it. The essential contours of the adversarial relationship remain unchanged. In America, the ground between capture and coercion remains narrow.

# CHAPTER SEVEN

# Government Regulation and Comparative Politics

Chapter 6 suggested that each nation's approach to the regulation of corporate social conduct was rooted in the history of its pattern of interaction among business, government, and the public. This explanation, however, by no means exhausts the significance of the contrasts in regulatory styles we have observed. For business–government relations do not take place in a vacuum; they constitute one dimension, albeit an extremely important one, of a nation's public policy processes. To understand fully each nation's approach to the regulation of industry, we need to examine its links to its constitutional system as well as its political culture.

## COMPARATIVE POLITICAL ECONOMY

Over the last thirty years, various efforts have been made to classify the patterns of interaction between business and government in major industrial societies. Among the earliest and perhaps the best known is that of Alexander Gerschenkron. He notes that

> in a number of important historical instances industrialization processes, when launched at length in a backward country, showed considerable differences, as compared with more advanced countries, not only with regard to the speed of the development (the rate of industrial growth) but also with regard to the productive and organizational structure of industry which emerged from those processes.[1]

The later a nation industrializes, Gerschenkron argued, the larger and more important is the role of its government in accumulating and allocating capital. Early industrializers tend to have well-developed capital markets and relatively autonomous firms, while in those nations that industrialize later, firms are more likely to be dependent on either banks or the state as a source of capital. Writing a little more than a decade later, Andrew Shonfield also placed the United States and the nations of Western Europe in two broad categories: nations in which the government plays an active, legitimate, and effective role in seeking to coordinate economic decision making and those in which the relationship between business and government remains at arm's length.[2] In 1974 Philippe Schmitter identified two principal patterns of interaction between government and interest groups among the capitalist democracies: pluralism and corporatism.[3] The distinction between them revolves primarily around the extent to which the activities of pressure groups are structured and officially recognized by the state.

Writing on American commercial and monetary policy in the mid-1970s, Stephen Krasner suggested that "the defining characteristic of a political system is the power of the state in relation to its own society."[4] Accordingly, he distinguished between "weak" states—those unable to formulate policy goals independent of particular groups in their society, to change the behavior of specific groups, or to alter directly the structure of their society—and strong states, which are able to accomplish each of these objectives. Peter Katzenstein divides the dominant political forms of contemporary capitalism into three categories: liberal, statist, and corporatist.[5] John Zysman has also advanced a tripartite classification, distinguishing "three distinct technical-political solutions to the problems of growth" and industrial adjustment: state-led, market-led, and negotiated.[6] The approach a nation adopts, Zysman contends, is determined largely by the structure of relationships among its financial institutions, government, and industry.

These efforts and others have stimulated a lively debate among students of comparative political economy. Scholars have disagreed both about the criteria used to define these categories and about the categories in which particular nations should be placed. On one issue, however, there does appear to be consensus: whatever classification scheme is employed, Great Britain and the United States are invariably placed in the same category. Thus for Gerschenkron the two Anglo-Saxon democracies are both "early industrializers," while for Shonfield they are the two capitalist nations in which business and government have experienced the greatest difficulty in coordinating their economic policies. For Schmitter, Great Britain and the United States are among the industrial nations that "have proven consistently more resistant to the blandishments of corporatism."[7] While Krasner does not explicitly refer to any nation other than the United States in his essay, given his criteria, Britain, like the United States, would logically be classified as a weak state. Katzenstein explicitly classifies both Great Britain and the United States as liberal states. Zysman argues that the pattern of industrial adjustment of the United States has been market-led. But while he is unable to place Britain in one of his three categories, on the basis of his independent variable—namely, the way in which a nation's financial markets and organized—Great Britain and the United States exhibit similar characteristics: they are the only two capitalist nations in which private equity markets are the principal mechanisms for allocation of capital. Accordingly, both states lack sufficient control over their financial institutions to direct the pattern of industrial adjustment.

With the exception of Schmitter, each of these analysts was concerned primarily with explaining various dimensions of foreign and domestic economic policy. None explicitly dealt with the dimension of government–business relations addressed in this study, namely, government regulation. In this policy area, however, it is the differences between Britain and America, not the similarities, that are most striking. Both Great Britain and the United States may be early industrializers, but they have come to pursue markedly divergent strategies for controlling the social behavior of their industries. Shonfield's description of the United States as a nation in which the role of public power is "uncertain" may help explain its inability to develop a coherent industrial policy, but it certainly is less useful as an explanation of the size, scope, and power of its regulatory bureaucracy. Nor is his characterization of the relationship between business and government in Britain as "arm's length" consistent with the widespread cooperation that exists between civil servants and industrialists across a wide range of areas of regulatory policy.

America, in Krasner's formulation, may be a weak state when measured by its government's ability to reshape the structure of its economy, but its efforts to control corporate social conduct suggest a state that has been relatively successful in challenging the prerogatives of industry. In neither Britain nor the United States can the pattern of interaction among industry, organized labor, and the government be classified as corporatist. Nonetheless, in many areas of British regulatory policy the relationship between private organizations and the government does approximate a corporatist one. And the failures of Britain's policies to improve the nation's economic performance, noted by both Shonfield and Zysman, stand in sharp contrast to its accomplishments in controlling the externalities produced by industrial growth.

It is not only that Great Britain and the United States diverge far more sharply from one another in the making and implementation of regulatory policy than the comparative political economy literature would lead one to expect. The magnitude of this divergence is even more striking. For if one were to construct a typology of national approaches to regulation, on various dimensions, Britain and America would fall at opposite ends of the continuum.

The uniqueness of the American approach to regulation is the one finding on which every cross-national study of regulation is in agreement. The American system of regulation is distinctive in the degree of oversight exercised by the judiciary and the national legislature, in the formality of its rulemaking and enforcement process, in its reliance on prosecution, in the amount of information made available to the public, and in the extent of the opportunities provided for participation by nonindustry constituencies. Although there are some exceptions—Sweden's consumer protection standards, including its regulations governing the approval of new ethical drugs, are stricter than those in the United States, and Japan has imposed tighter restrictions on its bioengineering industry—on balance the restrictions the United States has placed on corporate conduct affecting public health, safety, and amenity are at least as strict as and in many cases stricter than those adopted by other capitalist nations. As a result, in no other nation have the relations between the regulated and the regulators been so consistently or so persistently strained.

Students of comparative government regulation have been less sensitive, however, to the uniqueness of British regulatory practices. This shortcoming has been particularly noticeable in cross-national studies by American scholars. Both Joseph Badaracco's analysis of the regulation of the production of vinyl chloride in the United States, Great Britain, France, and Japan and the study of chemical regulation in the United States, Great Britain, France, and the Federal Republic of Germany by Ronald Brickman and his colleagues distinguish two approaches to regulation: the one adopted by America and the one adopted by the other countries in their sample.[8] This dichotomy certainly is a useful one: by international standards, the American approach to regulation is indeed distinctive. But what these authors have overlooked is that the British system of regulation is also unique: it differs as much from that of other industrialized nations as does that of the United States. We can, in fact, distinguish roughly three broad national patterns of government regulation: The British, the continental, and the American.

Within Europe, the British approach, with its common law orientation, differs sharply from the legal orientation of the continental nations, with their roots in the civil (Roman) law. The continental approach differs from that of Britain in its emphasis on precise rules and timetables and centrally determined standards; in virtually every policy area, the continental system of regulation is more formal and rule-oriented than that employed by the British. The latter's regulatory policies have also changed more slowly; their goals have remained incremental. Regulatory authorities on the Continent have also consistently established stricter controls over potential hazards than their counterparts in Britain. On each of these dimensions the continental approach to regulation has more in common with that adopted by the United States than it does with Britain's.

The British system of regulation also differs from the continental one in respect to the relationship among interest groups, government, and the public. The European system of regulation may be a relatively closed one in comparison with that of the United States; but in important respects the British system is far more open than that of other European nations. Britain makes more extensive use of nongovernmental organizations to implement regulatory policies and more closely consults with a wide variety of nonindustry interest groups before specific policies are formulated. The British system of land-use planning in particular provides both amenity and community groups with far more opportunity to influence planning decisions than the system employed in both France and the Federal Republic of Germany.

There is thus no necessary relationship between the organization of a nation's political economy and its approach to government regulation: Britain and the United States are roughly similar with respect to the former but very different with respect to the latter. This does not mean that we must abandon the categories developed by students of comparative political economy; rather we need to apply them with greater precision. It is important to distinguish different patterns of policy making not simply among nations but within them. We need to investigate, for example, which policies a nation makes through mechanisms of interest-group participation that are corporatist and which it makes through mechanisms that are pluralist: a nation may employ corporatist modes of interest-group representation with respect to one set of policies affecting industry but may resolve other dimensions of business–government relations through modes of interaction that are essentially pluralist. Similarly, instead of seeking to evaluate the relative "strengths" or "weaknesses" of particular states, we need to specify those policy areas in which government seems to be capable of asserting its prerogatives vis-à-vis interest groups and those areas in which it is dominated by them. For, as Zysman notes, "a government's ability to act in one policy area will be very different from its ability to act in another. . . . The policy tasks in each sector vary, as does the pattern of interest organization."[9] In short, we need to apply the concepts of comparative political economy to the analysis of government regulation.

## THE BRITISH POLICY STYLE

Samuel Beer has noted the widely shared attitude on the part of both policy makers and the public that

> organized groups have a "right" to take part in making policy related to their sector of activity; indeed that their approval of a relevant policy or program is a substantial reason for public confidence in it and conversely that their disapproval is cause for public uneasiness.[10]

In his study of pressure-group politics in Britain, Harry Eckstein remarked on the persistence of "corporatist attitudes" in a large number of policy areas, and hence

> the prerequisite normative insistence on negotiations between government and voluntary associations on matters of policy. . . . In Britain, at any rate, a policy regulating, say, farmers, embarked upon without close conversations between government and farm organizations, would be considered to be only on the margins of legitimacy, whether highly technical in character or not. . . . It is in short an attitude reflecting the widespread acceptance of functional representation in British political culture.[11]

Nearly two decades later Grant Jordan and Jeremy Richardson suggested that the style of policy making in Britain is characterized by five overlapping elements: sectorization, clientelism, consultation/negotiation, the institutionalization of compromise, and the development of exchange relationships. They argue that each is rooted in the importance attached to consensus and a "desire to avoid the imposition of solutions on sections of society." They write: "Underlying the consultative/negotiative practice is a broad cultural norm that the governing should govern by consent."[12]

These norms are most obviously reflected in the relationship between regulatory officials and industry. As we have seen, policy makers in Britain work closely with the industries whose conduct they are responsible for supervising. They rely heavily on their expertise and advice, generally secure their consent before formulating changes in policies, and whenever possible rely on them to implement the rules and regulations that are then adopted. It is not just industry whose consent the

*269*

British government has attempted to secure. Representatives of labor must also give their approval before the Health and Safety Executive can promulgate rules and regulations governing health and safety at the workplace. Likewise, an important purpose of the Town and Country Planning system is to allow community residents affected by a proposed development to decide whether or not they wish to grant approval to it. While the central government overrides the preferences of local authorities on occasion, it does so relatively infrequently, and usually only when a planning decision has national implications. Even in the latter case, as the history of the construction of a third international airport to serve the London area and the effort to develop an energy facility at Drumbuie demonstrate, community groups have been able to frustrate national policies. Moreover, the British central government has never sought to deny planning permission for a development strongly supported by local residents. On balance, notwithstanding the fact that ultimate legal authority resides in the central government, local authorities continue to enjoy considerable discretion in implementing policies that affect both land use and pollution control.

Significantly, three of the most conspicuous failures of British environmental policy in the postwar period stem from the unwillingness of the central government to impose restraints on constituencies without their consent. And neither directly involves industry. Although given the power to do so by the Clean Air Act of 1967, the Department of the Environment has never required a local authority to prohibit the burning of coal for domestic heating, despite the obvious justification for such regulation on both health and amenity grounds. The reason for this reluctance is a simple one: such an order would run directly counter to the preferences of the residents of the mining communities whose health the government is trying to protect. In fact, a major source of opposition to the Clean Air Act (1956) stemmed from the central government's conception of its role as a protector of individual property rights against local authorities.

Similarly, the hesitancy with which the government has moved to impose planning controls over agricultural land use reflects not simply the political power of Britain's farmers but also the government's preference for a system of voluntary restraints over a system of mandatory controls. Likewise it is the central government's reluctance to challenge the prerogatives of local authorities that has frustrated the efforts of British amenity organizations to enforce tighter restrictions on land use in Britain's national parks: the DOE remains unwilling to limit the control of local authorities over the land within their jurisdictions. In the areas of both pollution control and agricultural land use the government has sought primarily to encourage changes in individual behavior by providing cash subsidies—an indication of its clear preference for a voluntary system of controls over a more coercive one.

The importance attached to the values of both consultation and compromise helps account for much of the relative stability of British environmental policy. In his comparative study of British and American foreign policy, Kenneth Waltz writes that

> the movement of policy [in Britain] is customarily steady but slow; governments promote compromise and contrive adjustments between interests in conflict rather than meeting problems head on; ministers calculate carefully before giving even the appearance of leading. Continuity is impressive. . . .[13]

Characteristically the two most important changes in British pollution-control policy in the postwar period, the Clean Air Act (1956) and the Water Act (1973), were enacted following extensive and time-consuming negotiations. Likewise the government reached its decision to ban the use of lead in petrol over a twelve-year period; the imposition of planning controls over agricultural land use is also likely to take place gradually. Similarly, the fact that in Britain, as Jack Hayward observed, "there are no explicit, overriding medium or long term objectives" and that "unplanned decision making is incremental" is reflected in the reluctance of the British government to formulate explicit environmental objectives.[14] The result, Jordan and Richardson conclude, is that policies in Britain tend to be reactive rather than anticipatory—an observation that is confirmed both by the British approach to risk assessment and by the government's unwillingness to require environmental impact assessment.

Clearly, Britain's policy style extends beyond government regulation of industry. David Kirp notes that British central government officials have a profound "lack of taste for directive behavior." He quotes one official: "Consultation is a way of life with us." Kirp adds:

> In such a "pluralistic, incremental, unsystematic, reactive" system, formal power counts for less than informal suasion. Bold initiatives are far less common than nuanced pressures on local authorities. . . . Occasions when a policy change . . . is required, not merely recommended, are rare. In a ministry described by the Organization for Economic Cooperation and Development (OECD) as "pragmatic, conservative and evolutionary, not theoretical, futurological, and revolutionary," few rules will be issued, except after the achievement of consensus.[15]

This analysis suggests that what distinguishes government regulation of industry in Britain is not so much its pro-industry bias—after all, in the final analysis, the interests of industry do not appear to carry much less weight in American regulatory policy—but rather its bias in favor of the status quo. As a general rule, the people most closely consulted about a particular policy are those recognized by the government as having a disproportionate stake in its outcome. In the cases of the control of pollution and the introduction of new drugs and other potentially dangerous products, these are apparently leaders of industry; they are the ones on whom the direct burdens of compliance falls. For the most part, the public's stake in these policies is viewed as much more indirect. The only two instances in which the representation of nonindustry constituencies is roughly equivalent to that of industry involve occupational health and safety and land-use planning. In both cases there do exist distinct and readily identifiable groups of individuals whose stake in these policy decisions is relatively direct, namely, workers, community residents, and conservation groups. Accordingly, mechanisms have been established to solicit their views.

The importance attached to consent in British public policy can also be seen in the public's response when that value has been violated. Two of the most bitter environmental conflicts in recent years have involved the construction of additional motorways and the appropriate weight of heavy lorries. In both cases, popular resentment stemmed from a perception on the part of residents of particular communities that they were not being consulted before decisions that had a direct impact on their daily lives were being made. Likewise, criticism of the inadequacy of pollution-control regulations has almost always centered on the effect of pollution on particular communities: hence the considerable controversy over the nuclear wastes generated at Windscale. (In this context, the controversy over the lead content of petrol was highly unusual.) In general, the planning situations that have aroused the greatest controversy are those that have pitted the residents of individual communities against developments supported by the central government. Conflict in such cases is particularly bitter because the land suitable for some of these developments is so limited that both consultation and compromise are extremely difficult.

Indeed, what is striking about the British political system is the extent to which both governmental and nongovernmental organizations have been left free to govern themselves—free from controls imposed by the central government. We can see this clearly in the case of regulatory policy. In most areas, policy is made by a relatively autonomous network of departments and interest groups, generally insulated from both parliamentary and ministerial scrutiny. If one turns to the pattern of government control over universities, the pattern is similar: except for approving their budgets, the central government has interfered remarkably little with their day-to-day operations. Likewise, until relatively recently the City was effectively self-regulating. Even British trade unions, until recently, experienced virtually no government controls over their internal operations. Britain's nationalized firms have also been relatively free from direct governmental intervention in their operations. For the most part, government supervision of their activities has involved primarily the imposition of financial controls—and even these controls have not been vigorously imposed until recently.

All of those phenomena demonstrate the "persistent corporatism" of British social attitudes. This approach is predicated on "a conception of society as consisting primarily not of individuals, but of sub-societies, groups having traditions, occupational and other characteristics in common." Accordingly, "where corporatistic attitudes persist, governments tend to be regarded not as sovereigns in the Austinian sense, but, in the pluralistic sense, as corporations among many other kinds of corporations."[16]

The term "corporatism" has been used by political scientists in various ways. In general, however, it refers to one of two phenomena: centralized bargaining among interest groups and the granting of semiofficial status to specific interest groups, which then assist the government to implement those policies that directly affect them. The former usage usually refers to the making and implementation of economic policy: in political systems that are corporatist, official representatives of labor and industry are responsible for formulating and enforcing policies across a whole spectrum of issues that bear on both the interests of their membership and the management of the economy. These issues frequently include prices, wages, labor, and social welfare expenditures. By this criterion Britain is certainly not corporatist: its efforts to coordinate public and private policies more effectively in each of those areas have foundered on the inability of either the CBI or the TUC to control its constituent bodies. If, however, "corporatism" is used to refer to the latter phenomenon, then in many policy areas—including but not confined to government regulation—the mode of interest-group representation and policy making in Britain does resemble a corporatist one.

The most obvious example of a corporatist mode of interest-group representation is in the area of occupational health and safety policy. The Health and Safety Commission is composed of representatives of government, industry, and labor, while the making of specific rules for each industry is legally delegated to the latter two constituencies, each of which must approve any regulation. Likewise, the individuals appointed to serve on the regional water authorities established in the early 1970s were chosen to represent both industry and local authorities, the two groups with the most direct stake in British water policy. A similar function is performed by the large number of advisory bodies that advise the government on virtually every aspect of regulation, particularly in such technical areas as drugs. Both the official and semiofficial nature of the role played by interest groups in Britain is particularly noticeable in the area of conservation policy, where a number of organizations are routinely consulted before policies affecting land use are made. The designation "Royal" in front of the names of both the Royal Society for the Protection of Birds and the Royal Society for Nature Conservation signifies that these organizations enjoy a quasi-official status (the National Trust also has a royal charter).

Moreover, the British government not only is a major source of financial support for environmental organizations but in fact has played a major role in their creation. The Building Conservation Trust and the Societies for the Protection of Scotland, England, and Wales were all created at the initiative of the government, and a grant from the Countryside Commission restructured the Standing Committee on National Parks. Likewise the Committee for Environmental Conservation can be regarded as a peak organization of environmental groups. Composed of representatives of Britain's major environmental organizations, CoEnCo was established, in part, to enable the British environmental movement to speak with one voice in its negotiations with the government. The organization has subsequently been used by the government to solicit the views of British amenity organizations on a wide variety of policies within its area of expertise.

The British government also makes extensive use of nongovernmental organizations to assist it in policy implementation. The Pesticides Safety Precautions Scheme, for example, is based on a formal agreement between various government departments and industrial associations by which the latter have accepted exclusive responsibility for enforcing the regulations governing the approval and use of pesticides. Compared to other capitalist nations, Britain makes unusually extensive use of nongovernmental organizations to assist it in implementing its conservation policies; only a small fraction of the land whose use is restricted for conservation purposes is owned by the government itself.

Claus Offe suggests that an important purpose of the corporatist mechanism of interest-group representation is to help the government to limit the number and intensity of the demands made upon it and thus to minimize the scope of political conflict. He writes:

> In a typical case access to government decision making positions is facilitated through the political recognition of an interest group, but the organization in question becomes subject to more or less formalized obligations, for example to behave responsibly and predictably and to refrain from any nonnegotiable demands or nonacceptable tactics.[17]

As my description of the dynamics of British regulatory policy suggests, the British government is by no means a passive receptor of political pressures exerted on it. Quite the contrary: it has at its disposal various mechanisms both to limit and to channel interest-group participation. Participation in the making of most regulatory policies, including pollution control and the regulation of potentially dangerous substances, is effectively limited to representatives of industry. Other constituencies are excluded. At the same time the extensive involvement of trade associations in these policy areas does serve to encourage them to behave "responsibly." In fact, it is common practice for regulatory officials to appeal to either a trade association or the largest company in a particular industry to put pressure on those firms that are behaving inappropriately.

The participation of trade unions in occupational health and safety regulation serves a similar function: it both limits the scope of political conflict to a particular forum and ensures the support of both industry and labor for rules and regulations once they have been made. Much the same may be said of the discretion British civil servants enjoy in determining which groups they should consult: in effect, the granting of consultation status confers a de facto monopoly of representation.

> Groups have trade(d) off certain rights to frustrate the government by all possible means—in return for the productivity, the insurance of consultative status, and the "standing" in the policy-making community that inside status confers . . . in the pressure of a group/government nexus, a stress on conflict-avoidance has developed.[18]

J. J. Richardson and A. G. Jordan observe that "if the late Martin Luther King, Jr., had lived in London rather than Montgomery, Alabama, he would not have been the moving force behind a revolution in race relations. He would have been chairman of the Community Relations Commission."[19]

Even in political systems in which corporatist mechanisms of interest-group representation are highly developed, conflicts periodically emerge that cannot be satisfactorily mediated by such groups. This has certainly been the case in Britain. Not all regulatory policies have been capable of being resolved through private negotiations among interest groups or between interest groups and civil servants. From time to time they spill over into the broader political system. In such cases the roles of both Parliament and the cabinet become much more important. Chapter 1 and Chapter 3 discussed several such cases: mining in Snowdonia National Park, the construction of an energy-related facility at Drumbuie, the lead content of petrol, the siting of a third international London airport, and the control of poisonous wastes. In each case the government was clearly unsuccessful in limiting either the scope of political conflict or the number of political participants. The resulting pattern of interest-group participation can be described more accurately as pluralist than as corporatist. In the controversies over airport construction and the lead content of petrol, new organizations were expressly formed to deal with these issues and then disbanded when they had achieved their objectives. Similarly, the political style originally adopted by the Friends of the Earth was explicitly designed to undermine the Whitehall-based consensus on which previous British environmental policy had been based. More generally, the emergence of the large public inquiry can be seen as a failure of corporatism: participation tends to be relatively unstructured, and the inquiries end not in policies based on consensus but with clearly identifiable winners and losers.

 Nevertheless, these examples of pluralist politics remain exceptional: compared with those of the United States, a relatively small portion of British regulatory policies have become objects of public debate and conflict. Despite the increased willingness of British community and environmental organizations to challenge government policies publicly, most continue to rely on what are essentially "insider" political strategies. In their study of British pressure-group activity, which includes an extensive discussion of British environmental policy, Richardson and Jordan conclude:

> The Whitehall preference in the processing of policies is to internalise the required debate within some structure or institution. . . . Of course, processing of problems in such a fashion does not mean the resolution of all issues. Some obviously spill over into the party/parliamentary/political arena. [In general] both sides try to reach agreement *before* issues go to Parliament for formal ratification. So the cases where Parliament is called in to resolve a conflict are comparatively rare. These are instances when the normal processing system has "blown a fuse."[20]

## THE U.S. POLICY STYLE

While America has rarely been described as corporatist, various studies of the role of particular interest groups in the political process have suggested that many areas of public policy in the United States do have a distinctly corporatist flavor. Grant McConnell writes:

> A large number of groups have achieved substantial autonomy for themselves and the isolation of important segments of government and public policy. The result has been the establishment of varying degrees of control and exercise of public authority by the private groups within the public areas with which they are concerned.[21]

Theodore Lowi's concept of "interest group liberalism" denotes a similar phenomenon: the parceling out of state sovereignty among particular interest groups, each of which is then in a position to exercise quasi-public authority.[22] A number of studies of the relationship between various regulatory agencies and the industries whose conduct they are responsible for controlling reach a similar conclusion: the dynamics of industry capture suggest a pattern of interaction between specific industries and various government agencies that is corporatist in all but name.[23] In addition, the fragmentation of public policy described by McConnell appears quite similar to the "departmental pluralism" noted by students of British politics. The substantial discretion accorded to administrative agencies and the frequent use of nongovernmental organizations to implement public policies—both described and analyzed by Lowi in *The End of Liberalism*—as well as the restrictions on nonindustry access to the regulatory processes documented by capture theorists of government regulation, all suggest important similarities between the British and American public processes. Clearly in both countries, as Eckstein has said of Britain, "there is a very close and continuous collaboration between pressure groups and [government officials]: so close that in some instances groups affected by departmental activities have been almost directly assimilated into the Department."[24]

To the extent that these quasi-corporatist mechanisms of interest-group representation are typical of the American public policy process, the relationships between interest groups and the state in both Britain and the United States appear to be similar. These similarities, however, are superficial. The pattern of interest-group representation has been consistently more pluralist in the United States than in Britain. Significantly, most of the literature suggesting that the American system of interest-group representation exhibited corporatist tendencies was written during the 1950s and 1960s, a period when many aspects of American domestic policy were in fact relatively stable. This was certainly true of government regulation of business. But however accurately Lowi and McConnell may have characterized the period in which they wrote, their analyses, like those of the capture theorists of regulation, have been rendered obsolete by events of the last two decades. Not only with respect to government regulation but in a wide variety of policy areas American politics has become significantly more pluralist. Far fewer decisions are resolved and implemented among an "iron triangle" of a congressional committee, an interest group, and a government agency than was the case two decades ago. The significant increase in the number of nonbusiness interest groups since the mid-1960s has made the making of public policies affecting business more public—and their outcomes more problematic.[25] Likewise, fewer government departments are so closely identified with particular interest groups as was the case during the 1950s and 1960s; rather a far higher proportion have become the objects of contention among different interest groups.

America's politics has always been more pluralist than Britain's. America, as a whole, has always been less well organized. The United States has, for example, no counterpart to the Trades Union Council, the Confederation of British Industry, or the National Farmers Union, which represent a significant portion of British workers, businesses, and farmers, respectively, and thus are in a position to negotiate on their behalf with respect to a wide variety of policies. Moreover, as Robert Salisbury observes, "there is virtually no official incorporation of formal associations as participants in the policy discussions. They are not invited to designate sector representatives on government advisory boards, or to name key policy-makers who are to deal with their sector."[26] Significantly, there is no American equivalent of CoEnCo, formed in the early 1970s as a peak organization of British environmental groups; American environmental organizations do work together, but only on an ad hoc basis. American environmental groups, unlike their British counterparts, lack any quasi-official status and are rarely employed by the government to assist it in policy implementation. They continue to pursue what are essentially "outsider" political strategies. Nor are trade unions formally consulted with respect to the making and enforcement of regulations governing health and safety at the workplace, as they are in Britain. Similarly, American trade associations have traditionally been far weaker than their British counterparts and only rarely have been granted the right to regulate the activities of their members; American regulatory policy has always relied far less on industry self-regulation than has Britain's.

The sources of this persistent pluralism are both structural and ideological. The American system of government, in sharp contrast to the British, provides interest groups with multiple sources of access. Because of the separation of powers, if pressure groups in America are not able to work out an accommodation with a particular department or agency or commission, they have numerous other political mechanisms at their disposal: they can turn to Congress, the courts, or, in many cases, the White House or the executive office of the president. Moreover, the fact that the senior policy-making officials in the United States maintain their positions for only a relatively short period of time—whereas in Great Britain such officials are likely to be civil servants—further diminishes the incentives of interest groups to develop close, permanent ties with government departments. They can always wait until the next election in anticipation that a new administration will be more sympathetic to their interests.[27]

Equally important, there does exist in America a "profound prejudice against corporate politics, against the organization of opinion by 'interest' groups."[28] The notion that the relationship between public officials and lobbyists should be "free, easy, open and intimate" goes against Americans' traditional mistrust of the role of interest groups in the political process—a prejudice that dates back to *The Federalist Papers* and is reflected in the "heirs of the progressive tradition [who] characteristically regarded interest groups as the enemy of the public interest."[29] Accordingly, "an important reason the American public policy has not enhanced corporatist tendencies is that monopolistic interest groups are regarded with deep suspicion in the American political culture."[30] This mistrust is particularly pronounced when a particular interest group is perceived as enjoying a privileged position, as appeared to be the case with industry in many policy areas during the 1950s and 1960s.

If quasi-corporatist modes of interest-group representation reflect an important strain in British political culture, then American political culture is similarly biased toward pluralism—open, unstructured competition among interest groups. In this context, what is striking about both Lowi's and McConnell's analyses is their ethnocentrism. Their criticisms of the American public policy process are themselves reflections of American political values. Specifically, they criticize as illegitimate precisely those aspects of the relationship between government and interest groups that students of British politics have traditionally regarded as a source of the legitimacy of the British system of governance. Only in America are quasi-corporatist ties between interest groups and the state viewed as undermining rather than strengthening the legitimacy of public authority.

Quasi-corporatist arguments require that administrators be given substantial discretion; otherwise they will not be able to commit the authority of the government in their negotiations with particular interest groups.

279

> Negotiations . . . demand the concentration of authority on both sides, as well as the vesting of considerable discretionary authority in the negotiators. Indeed the latter presupposes the former. Genuine negotiations between governmental bodies and pressure groups are not likely to take place when a decision must be obtained from a large number of bodies before it has force—as in the American separation-of-powers system.[31]

Indeed, the argument that the contemporary American public policy process has substituted the authority of administrators for the rule of law is central to Lowi's critique of it. But while, as I suggested in Chapter 6, the amount of discretionary authority extended to administrative agencies in the United States has varied over the last two centuries, on balance American officials enjoy less discretion than administrators in any other democracy; in no other nation are they so accountable to both the courts and the legislature. In fact, American political culture "has always contained a powerful strain of distrust of governmental officials." Michael Asimov writes that this distrust of bureaucracy

> underlies the prevailing view that, whenever possible . . . broad discretion should be fettered, or at least structured, i.e., if not by statutory standards, then by regulations, or by self-imposed, nondelegated subordinate legislation (such as guidelines or policy statements), by adherence of precedent or through formalizing the decisional process.[32]

One of the most important ways in which the autonomy of administrative officials has been restricted in the United States has been by denying them control over access. Officials in America cannot decide which interest groups they wish to allow to testify at hearings, to receive documents, or to disclose information to; in principle all interest groups must be treated equally. The traditional American reliance on litigation as a means of resolving public policy issues can be seen in similar terms: the courts represent the one arena of government to which relatively equal access can be most readily granted. Over the last decade the judiciary has both itself progressively lowered the barriers to politically motivated litigants and made administrative agencies adhere to more rigorous standards of procedural "fairness." Moreover, the federal government has, both indirectly and directly, become an important source of financial support for challenges to the discretion of regulatory agencies. Characteristically, the most important way in which environmental groups in America "assist" regulatory agencies in policy implementation is by suing them.

## GOVERNMENTAL REGULATION AND ECONOMIC PERFORMANCE

In comparison with the policies of other industrialized democracies, British regulatory policy in general and environmental regulation in particular must be judged successful. Its success can be measured not only by its progress in achieving its professed objectives but also by the relatively high degree of legitimacy associated with it. Unlike, for example, Japan, which did not begin a serious effort to enact environmental controls over industry until it was confronted with widespread evidence of death and disease, Britain has been attempting to protect its people, wildlife, and land from the hazards of industrial production for well over a century: regulations protecting the health and safety of workers were initially established early in the nineteenth century, and the Alkali Inspectorate, established in 1861, is the world's oldest pollution-control agency. Britain appears to have made a more extensive effort to balance amenity and economic values than any other nation in the quarter-century following World War II. Its system of land-use planning is more extensive than that of any other country, and was well in place considerably before the upsurge of public interest in environmental quality that occurred in every industrialized nation in the late 1960s.

Over the last fifteen years, Britain's efforts at social regulation have not been noticeably more effective than those of other nations; at least in the area of environmental policy, the nation that has expended the most resources and made the most progress is Japan.[33] On the other hand, Britain's occupational injury rate is lower than that of either France or the Federal Republic of Germany and during the 1970s declined more than that of the United States. Its coal mines are also considerably safer than those of any other major coal-producing nation.[34] Its age-adjusted rate of cancer deaths is lower than those of Belgium, the Netherlands, Austria, and France, although it is higher than those of West Germany, Italy, Japan, Sweden, and the United States; the average life expectancy of its population is lower than that of the United States, Sweden, Japan, and France, but greater than that of West Germany. Unfortunately, other indices of regulatory effectiveness are scarce. There is a second yardstick, however, by which we can measure policy success: a political one.

Britain's achievement lies in the ability of its political system to adjust to the public's increased interest in health, safety, and amenity issues and to increased demands for participation in the regulatory process without undermining the close ties with industry on which its compliance strategy has historically rested. (Indeed, for all of Britain's uneasiness with the efforts of the European Economic Community to restructure its regulatory rules and procedures, it is striking how little the EEC has really affected the relationship between regulatory officials and industry in Britain.) The American political system has clearly been more responsive to the demands of nonindustry constituencies, but it has paid a heavy price: a dramatic increase in conflict and mistrust between government and industry. West Germany, France, and Japan have been able to maintain relatively close and cooperative ties between regulatory officials and industry, but they have been relatively unsuccessful in integrating nonindustry constituencies into the policy process.

This analysis has repeatedly contrasted the stability of the British system of regulation with that of the United States. But when viewed in their own terms, Britain's institutions have demonstrated a continual ability to adjust to changing public concerns and expectations. During the 1940s Britain introduced a comprehensive system of land-use controls, which in turn became a vehicle for increased public participation in the shaping of controls over industrial location during the late 1960s. During the 1950s, in response to heightened public concern over the problem of air pollution, the government significantly expanded the jurisdiction of its ninety-year-old pollution-control agency, the Alkali Inspectorate, and increased the power of local authorities to control air pollution. In 1973 it instituted a fundamental reorganization of its system for controlling water pollution. It has also progressively increased the amount of information made available to the public. Even the emergence of the large public inquiry, for all its inadequacies and limitations, can be viewed as an attempt on the part of the government to provide a forum for the public discussion of important environmental policies. The two-decade delay in securing a site for a third international airport to serve the London area can be regarded as a policy failure, but from another perspective it also demonstrates the relative accountability of British planning procedures. In both Japan and West Germany, where public access to planning decisions is far more restricted, the battle over new airport construction has been much more bitter and frequently violent. In addition, citizen groups opposed to nuclear power have had far less effect on policy in France and West Germany than in Britain; hence the latter's still greater reliance on fossil fuels as a source of power generation. In comparison with that of the United States, the British system of regulation may be a relatively closed one. But it is far more open than those of other capitalist democracies.

These reforms have played an important role in maintaining the legitimacy of the British system of regulation. It is of course difficult to disentangle the legitimacy of government regulation from that of the political system as a whole. It can be argued, for example, that the deference shown by the British public towards its regulatory officials is simply a reflection of its more generalized acceptance of authority. But in fact the attitudes of the British public toward authority have by no means been consistently deferential: certainly segments of the British working class have not been passive in their response to governmental policies that threaten their interests. And even in the area of regulatory policy, the public has on occasion become highly mobilized.

One admittedly crude measure of the attitudes of the British public toward the nation's system of environmental control may be found in poll data. According to a 1978 survey, the percentage of the British public who believe that "protecting the environment is very important and who are dissatisfied with what is being done about it at present" was 30 percent. In Italy the comparable figure was 64 percent, in France 53 percent, in Denmark 51 percent, in Belgium 49 percent, in the Netherlands 40 percent, in West Germany 40 percent, and in the Republic of Ireland 30 percent.[35] A more important index is the degree of strain that public concern about environmental issues has placed on the political system as a whole. Here the contrast between Great Britain on one hand and the Federal Republic of Germany and Japan on the other is marked. In both of the latter two nations the politics of environmental regulation has been relatively disruptive. The rise of the Green party in Germany can be traced in part to the unresponsiveness of the German government to the public's concern about the adequacy of its environmental regulations, and the residents of particular communities in Japan have been extremely bitter about the unresponsiveness of the ruling Liberal Democratic party to their grievances against the companies whose emissions have been injuring them.[36] In both nations environmental policy has been far more contentious than in Britain. The Japanese approach to promoting industrial growth may be a model for other capitalist nations, but the same certainly cannot be said for the way it has gone about controlling the externalities produced by that growth. On the other hand, the relative accountability of American environmental policy to nonindustry constituencies does not appear to have contributed to its legitimacy: America's environmental regulation may be as effective as Britain's, but its citizens certainly do not perceive it as effective. Dissatisfaction with environmental regulation is ubiquitous in the United States.

The relative success of British controls over industry stands in sharp contrast to the government's failures in a wide variety of other policy areas, most notably in its management of the economy itself. (It has been remarked—only half facetiously—that the Clean Air Act and the Falklands War constitute Britain's most notable policy achievements of the postwar period.) By any index, Britain's industrial and economic policies have been less successful than those of any other capitalist nations over the last four decades. Why have Britain's efforts to improve the international competitiveness of its industry been so much less successful than its efforts to control pollution or protect the safety of its workers?

Part of the explanation has to do with the relative competence of the civil service in the two policy areas. The policy areas described in this study tend for the most part to be relatively technical; in each the British civil service includes individuals with particular expertise, often accumulated as a result of their previous experience in industry. Not only do they operate from a set of clearly defined professional norms, but they are working in policy areas in which the British government has been extremely active for a considerable period of time—in some cases for more than a century. They thus are able to draw upon an accumulation of expertise and experience. The British civil servants responsible for shaping the government's economic and industrial policies, by contrast, tend to be generalists; they are far less likely to have had prior experience working in industry and they are unable to draw upon a considerable body of expertise. Not only have many of the governmental bodies for which they work been established relatively recently but they have continually been reorganized and restructured.

This, however, is only part of the explanation. For if the shortcomings of Britain's policies toward industry were due to either the training of its civil servants or their lack of prior experience in industry, improving Britain's industrial policies would be far less difficult than it has proved to be. Even when the government has drawn more heavily on people with business experience, its policy effectiveness has remained disappointing. And if Britain's failures were due to the relative inexperience of its officials in these policy areas, then one would expect their effectiveness to improve over time—which does not appear to be the case.

The effectiveness of British regulatory policy and the failures of Britain's industrial policies are related. The very style of policy making that improves the effectiveness of the former diminishes the effectiveness of the latter. The government is unable to implement policies in either area without securing the consent and cooperation of the companies affected by their decisions: both sets of policies have been, in practice, jointly administered by civil servants and industrialists. In neither case has the government been able to impose its priorities on industry; rather it has continually adjusted its policies to industry's own definition of its needs and capacities. In this sense, the thirty-nine sectoral working parties established in 1979 to implement the government's "new industrial strategy" are analogous to the trade associations that work with the Alkali

Inspectorate to define the best practicable means of abatement for their industries. The only difference between the National Enterprise Board and the Alkali Inspectorate is that the former allocates money, the latter advice. Peter Hall writes:

> The striking feature of British intervention is its reliance on . . . schemes that are essentially still directed from the private sector. . . . Industries were essentially asked to rationalize themselves. . . . Such a consensual approach stands in striking contrast to the more *dirigiste* policies of France or Japan.[37]

A regulatory policy that operates by consensus does have certain limitations; it makes it more difficult for the government to respond quickly to force major changes in corporate practices. And an industrial policy that is insensitive to the preferences of industry may produce decisions that are impossible to implement. The former is apt to be too rigid, the latter ineffective. But on balance, what is a virtue in one policy area is likely to be a disaster in the other. It is one thing to work with existing business firms to encourage them to bring about relatively modest changes in the way they control emissions, test new drugs, or protect their workers. As we have seen, such a strategy has proved, on the whole, relatively effective. But the purposes of industrial policy are quite different: they have to do primarily with reorganizing the structure of an existing industry either to improve its competitive position or to reduce the resources devoted to it. In either case, the status quo must be challenged: companies and industries must be required to make fundamental changes. But this is precisely what the British government has not been able to do. Largely proceeding on an ad hoc basis, it has been unable or unwilling to impose priorities on the private sector to force the pace of economic adjustment. As John Zysman writes:

> Interventionist policies run against the training and traditions of the bureaucracy; indeed, an instinctive resistance to the selective use of state power seems built into the norms of the British civil service. The principles of hands-off government have been embedded in a tradition of administrative law which enjoins bureaucrats from making selective decisions that discriminate against particular firms.[38]

A defining characteristic of British environmental policy in the areas of both pollution control and industrial siting has been the unwillingness of central governmental officials to impose stricter regulatory standards on communities than the individuals who live and work there prefer. During the 1960s and 1970s the British government sought to follow a similar policy in regard to the closing of noncompetitive industrial plants.

In this policy area, too, it has, at least until recently, hesitated to impose hardships on particular communities; hence the government's extensive subsidies to noncompetitive industries and plants in economically depressed regions. As a result, much of British industrial policy—and this has been particularly true of the government's regional policies—has consisted of little more than the indiscriminate issuing of checks. In effect, Britain's industrial policies, like its regulatory policies, have not been technology-forcing. The policy styles may be similar in the two cases, but their economic consequences differ markedly.

An analogous argument can be made about the role of British business culture in the two policy areas. The same set of values that permits and encourages companies to cooperate with each other in formulating and implementing industry-wide standards of corporate social conduct also serve to limit their competitiveness with each other. The cartel-like mentality of much of British industry serves Britain extremely well when it comes to enforcing regulatory policies; it has precisely the opposite effect, however, when it comes to improving the performance of its economy. (In this context, it is noteworthy that the decline of the effectiveness of self-regulation in the City can be attributed primarily to growing competition in British financial markets—brought about largely by the growth in the number of foreign firms with offices in London.)

Whether or not government regulation of industry in the United States can be judged a success depends on one's criterion. If one evaluates the success of a policy area by the scope and extent of political participation, then American regulatory policies have been extremely successful: no other nation has provided such a wide range of opportunities for its citizens to participate in the making and enforcement of its controls over industry. On the other hand, if one measures policy success by the ability of a political system to minimize the scope of political conflict, then government regulation in America has less to commend it: over the last fifteen years it has created more conflict than that of any other industrial society. And it has not been noticeably more effective.

No matter which criterion one employs, in neither country can regulatory policy and economic performance be viewed in isolation. America's highly adversarial system of regulation is the counterpart of a highly competitive economic system. The conflict-ridden relationship between business and government in the United States is the political counterpart of the highly competitive relationships that exist within the American business community. Ironically, the elaborate and complex set of rules and regulations with which America has enveloped its industry testifies to the latter's highly competitive nature: were American firms less aggressive, the American public probably would not insist that their conduct be so closely regulated. In short, just as, in some sense, Britain's dismal economic performance may be the price it pays for its relatively successful system of regulation, America's adversarial style of regulation may be the price it pays for the relative competitiveness of much of its industry.

## CONCLUSION

In a sense, each nation's constitutional system tends to obscure the reality of governmental decision making. Legally, the British government—that is, the Crown acting through Parliament—is more powerful than that of any other democracy. In reality, the relative power of the British state vis-à-vis British society is sharply limited. The American constitutional system—that is, federalism and the separation of powers—was designed to minimize the likelihood that government would exercise its power in a coercive manner. In fact, the fragmentation of political decision making in the United States has not diminished the ability of various interest groups to use the government as a vehicle for coercing other political factions. On the contrary, it has multiplied their opportunities to do so. In several critical respects the United States has more in common with France than with Great Britain; political decision making may be more fragmented in the United States than in France, but the government is capable of being just as coercive.

The difference is that the French government, like that of Japan, does have a variety of instruments at its disposal by which it can both shape the direction of industrial activity and minimize access to the political process by various nonindustry constituencies, particularly organized labor and middle-class pressure groups. The American government has far less ability to do either. Yet at the same time its very openness and decentralization make it possible for a wide variety of interest groups to capture a slice of public authority and thereby use the power of government to impose their values on other interest groups or institutions, including business, in much the same way that public bureaucracies operate in France and Japan.

The American experience in the area of social regulation demonstrates that a strong state does not require a powerful or independent bureaucracy. Nor need it restrict nonindustry constituencies' access to the political process. The American state is not weak because it is permeated by interest groups. The opposite is more nearly the case: it is precisely the vulnerability of American regulatory officials to nonindustry pressures that has provided them with the legal and political resources to mount such a spirited challenge to corporate prerogatives. Correspond-ingly, one reason why regulatory authorities in Britain can be said to be in a relatively weak bargaining position vis-à-vis industry is that they are relatively insulated from nonindustry pressures. In most policy areas, the British system of regulation prevents, or at least makes problematic, the mobilization of nonindustry constituencies capable of giving regu-latory officials the political backing they would need were they to seek to drive a harder bargain with industry.

Yet this analysis suggests the limitations of the weak/strong state di-chotomy. On one hand, the American state is stronger in that it appears more capable of imposing policies on industry without its consent; the widespread use of prosecution represents one index of this power. At the same time, this reliance on rule-oriented enforcement can also be viewed as a sign of weakness; it demonstrates the inability of government officials to rely on more informal methods of social control as a means of achieving compliance on the part of industry. In both cases, the ob-verse applies to Great Britain: its informal approach to enforcement and minimal use of prosecution can be viewed as signs of either its power over industry or its dependence on it.

Conversely, the British government certainly does appear to be in a stronger bargaining position vis-à-vis various nonindustry constituencies. It is stronger because, whereas in America political participation has been defined as a matter of right, in Great Britain, to a large extent, it remains a privilege. Indeed, one might well argue that an important source of the relative stability of the British political system has been precisely the historic ability of the government to demand "socialization" as the price for political participation. Thus the middle class was given the franchise only after it no longer appeared to threaten the prerog-atives of the landed aristocracy. Similarly, in the latter part of the nine-teenth century, the working class was gradually allowed to participate in the electoral process as it shed its radicalism. And environmental groups are more likely to receive consultation papers when they have demonstrated that they are prepared to behave "responsibly."

The notion that the American liberal tradition contains an important element of coercion was made repeatedly by radical political scientists during the 1960s. In fact, there is nothing particularly radical about this idea: seventeenth- and eighteenth-century liberal political theorists in both Britain and America were in favor of limited government, not weak government. Where these critics erred was in suggesting that the exercise of coercion affected only "progressive" social groups—Indians, blacks, workers. In fact, the American government has sought to coerce both the left and the right. The southern plantation aristocracy was destroyed by the federal government in the course of the most violent military conflict of the nineteenth century. Over the next seventy-five years American workers were subjected to more government-sanctioned violence than those of any other industrial society. The New Deal represented a more significant challenge to corporate prerogatives than that experienced by any other industrial society in the 1930s, with the exception of Germany. Since the late 1960s, governmental coercion has most commonly been exercised by the judiciary and has generally been initiated by left-of-center political forces. On balance, the overall thrust of federal regulation of corporate social conduct in the areas of environmental protection, consumer protection, equal employment, and occupational health and safety has involved the exercise of more coercion vis-à-vis industry than that of any other industrial society.

Nor is it simply industry whose autonomy is constrained by American statutes and the *Federal Register*. Ironically, despite the fact that the United States has a federal system of government and Britain a unitary one, the policies of state and local governments are far more closely regulated and controlled by the federal government than is the case in Britain: many of the recent complaints of local governments about federal regulatory policies echo those of business. While the central governments of both nations have, for example, occasionally required local governments to accept development projects opposed by the latter, only in America has the federal government attempted to deny approval for developments *favored* by local governmental units. The federal government also exercises far more control over local education than does the British central government: the imposition of school busing on local communities would be inconceivable in Britain. Although much of American higher education is nominally private, the federal government regulates colleges and universities far more closely than does the British central government. Similarly, the internal operations of trade unions are far more strictly controlled in the United States than in Britain.

The similarity between the American and continental systems of social regulation is extremely significant. Indeed, American public law now appears to have far more in common with the continental legal tradition than with the common law tradition of Great Britain from which it is derived. What is the *Federal Register*, with its thousands of pages of detailed rules and regulations, if not the American equivalent of the Code Napoleon? As one legal scholar notes:

Britain has fewer regulations and most of them are less important than in America. To use a gross method of comparison, the number of newly adopted British statutory instruments in 1977, 1978, and 1979 total 1,918,

1,621, and 1,770, respectively, a great many of which relate to matters dealt with at state or local levels in America. In the U.S., it is estimated that 7,000 legislative rules are promulgated annually *at the federal level*, of which 2,000 have a "significant impact" on regulated parties or on competition and more than 100 have major economic effects (an economic impact of $100 million or more). By comparison, one who scrutinizes the annual harvest of the statutory instruments in Britain cannot fail to be impressed by the triviality of the vast majority.[39]

These fundamental differences in the nature of British and American politics are reflected in the way the term "pluralism" has been used by British and American political scientists. The term itself, as originally employed in Great Britain around the turn of the century, referred to a society made up of a plurality of institutions, one of which was the government. Ideally, each institution was to be relatively autonomous in its own sphere, thus minimizing the use of coercion by the government (the tradition of guild socialism derives from this conception). The classical liberal distinction between the public and the private spheres would likewise become obsolete. This vision bears a striking resemblance to the way the British political system actually works, with institutional boundaries and prerogatives the constant subjects of bargaining among private and public organizations. The term "pluralism" as it is used by political scientists on this side of the Atlantic refers to an entirely different phenomenon: the competition among private interest groups to use the power of public authority to further their own objectives, many of which are likely to challenge the prerogatives of other interest groups and thus other parts of government. And this appears to be what much of American politics—as well as much government regulation—is actually about.

# Notes

## Preface

1. The most important and widely cited is Louis Hartz's *The Liberal Tradition in America* (New York: Harcourt, Brace & World, 1955). Samuel P. Huntington, *American Politics and the Promise of Disharmony* (Cambridge: Harvard University Press, 1981), implicitly addresses some of the issues raised here.

2. See, for example, W. M. Wardell, "Introduction of New Therapeutic Drugs in the United States and Great Britain: An International Comparison," *Clinical Pharmacology and Therapeutics* 14 (1973): 773–92, and "Developments since 1971 in the Patterns of Introduction of New Therapeutic Drugs in the United States and Britain," in *Drug Development and Marketing*, ed. Robert B. Helms (Washington, D.C.: American Enterprise Institute, 1975), pp. 165–81; William Wardell, ed., *Controlling the Use of Therapeutic Drugs* (Washington, D.C.: American Enterprise Institute, 1978).

3. See, for example, Timothy O'Riordan, "The Role of Environmental Quality Objectives in the Politics of Pollution Control," in *Progress in Resource Management and Environmental Planning*, vol. 1, ed. Timothy O'Riordan and Ralph C. d'Arge (New York: Wiley, 1979), pp. 221–58; and O'Riordan, *Environmentalism* (London: Pion, 1980); Francis Sandbach, "A Further Look at the Environment as a Political Issue," *International Journal of Environmental Studies* 12 (1978): 99–113.

4. For example, Arnold Heildenheimer, Hugh Hector, and Carolyn Teich Adams, *Comparative Public Policy: The Politics of Social Choice in Europe and America* (New York: St. Martin's, 1983), contains ten chapters that focus on particular public policies; none deals with any of the issues addressed in this study.

5. Some exceptions include Michael Reich, "Environmental Policy and Japanese Society," pts. 1 and 2, *International Journal of Environmental Studies* 22 (1983): 191–98, 199–207; Carolyn Watkins, "Interest-Group Influence in Bureaucratic Decision Processes: The Implementation of Clean Air Policy in the U.S. and the F.R.G.," paper presented to the 1984 Annual Meeting of the American Political Science Association, Washington, D.C.; Michael Greve, "Institutional Determinants of Public Interest Politics: West Germany and the United States," Ph.D. dissertation abstract, Cornell University, 1984.

6. Cynthia Enloe, *The Politics of Pollution in a Comparative Perspective: Ecology and Power in Four Nations* (New York: David McKay, 1975).

7. Lennart Lundquist, *The Hare and the Tortoise: Clean Air Policies in the United States and Sweden* (Ann Arbor: University of Michigan Press, 1980); Steven Kelman, *Regulating America, Regulating Sweden: A Comparative Study of Occupational Safety and Health Policy* (Cambridge: MIT Press, 1981).

8. Peter Knoepfel, *Comparative Analysis of the Implementation of $SO_2$ Air Quality Control Policies in Europe: Conceptual Framework and First Results* (Berlin: International Institute for Environment and Society, Science Center, 1981); Peter Knoepfel and Helmut Weidner, "Formulation and Implementation of Air Quality Programmes: Patterns of Interest Consideration," *Policy and Politics* 10 (1982): 85–109, and "Implementing Air Quality Control Programs in Europe: Some Results of a Comparative Study," in *International Comparisons in Implementing Pollution Laws*, ed. Paul Downing and Kenneth Hanf (Boston: Kluwer-Nijhoff, 1983); Peter Knoepfel, Helmut Weidner, and Kenneth Hanf, *International Comparative Analysis of Program Formation and Implementation of $SO_2$ Air Pollution Control Policies in the EEC Countries and Switzerland: Analytical Framework and Research Guidelines for the National Research Teams* (Berlin: International Institute for Environment and Society, Science Center, June 1980).

9. Nicholas Ashford and George Henton, *Environmental Regulation of the Automobile* (Cambridge: Center for Policy Alternatives, MIT, 1982); Ronald Brickman, Sheila Jasanoff, and Thomas Ilgen, *Chemical Regulation and Cancer: A Cross-National Study of Policy and Politics* (Ithaca: Cornell University, Program on Science, Technology, and Society, 1982); Joanne Linnerooth, ed., *Risk Analysis and Decision Processes: The Study of Liquefied Energy Gas Facilities in Four Countries* (New York: Springer, 1983).

10. Joseph L. Badaracco, Jr., *Loading the Dice: A Five-Country Study of Vinyl Chloride Regulation* (Cambridge: Harvard Business School Press, 1985).

11. For a summary and critical analysis of the literature on cross-national environmental policy, see David Vogel, "The Comparative Study of Environmental Policy: A Review of the Literature," paper presented to the conference "Cross-National Policy Research: Lessons from Experience and Prospects for the Future," Science Center, Berlin, December 1983.

*Introduction*

1. Timothy O'Riordan, "Public Interest Environmental Groups in the United States and Britain," *American Studies* 13 (1979): 429–38.

2. See, for example, James Greene, *Regulatory Problems and the Perceptions of Business* (New York: Conference Board, 1982).

3. See, for example, Louis Hartz, *The Liberal Tradition in America* (New York: Harcourt, Brace & World, 1955).

4. Charles Lindblom, *Politics and Markets* (New York: Basic Books, 1977).

5. Andrew Shonfield, *Modern Capitalism* (New York: Oxford University Press, 1965), pp. 299, 325.

6. See, for example, R. Emmett Tyrell, Jr., ed., *The Future That Doesn't Work: Social Democracy's Failures in Britain* (Garden City, N.Y.: Doubleday, 1977); William Gwyn and Richard Rose, eds., *Britain: Progress and Decline* (New Orleans:

Tulane University Press, 1980); Stephen Young with A. V. Lowe, *Intervention in the Mixed Economy* (London: Croom Helm, 1974); Wyn Grant, *The Political Economy of Industrial Policy* (London: Butterworth, 1982); Trevor Smith, *The Politics of the Corporate Economy* (Oxford: Martin Robertson, 1979); Isaac Kramnick, ed., *Is Britain Dying?* (Ithaca: Cornell University Press, 1979). Strikingly, the one exception—Bernard Nossiter, *Britain, a Future That Works* (Boston: Houghton Mifflin, 1978)—does make a casual reference to Britain's achievements in the area of environmental policy; see pp. 69–70.

7. For the recent history of Japanese environmental policy, see Norie Huddle and Michael Reich, *Island of Dreams* (New York: Autumn Press, 1975); Margaret McKean, *Environmental Protest and Citizen Politics in Japan* (Berkeley: University of California Press, 1981); and T. J. Pempel, *Policy and Politics in Japan* (Philadelphia: Temple University Press, 1982), pp. 218–54.

8. See, for example, Shonfield, *Modern Capitalism*; Peter Katzenstein, ed., *Between Power and Plenty* (Madison: University of Wisconsin Press, 1978); Alexander Gerschenkron, "Economic Backwardness in Historical Perspective," in *Economic Backwardness in Historical Perspective*, ed. Gerschenkron (Cambridge: Harvard University Press, 1961); Philippe Schmitter and Gerhard Lehmbruch, eds., *Trends toward Capitalist Intermediation* (Beverly Hills, Calif.: Sage, 1979).

CHAPTER 1. *The Politics of Environmental Protection in Great Britain*

1. Eric Ashby and Mary Anderson, *The Politics of Clean Air* (Oxford: Clarendon, 1981), p. 82.

2. Quoted in Trevor Holloway, "The Restoration of the River Thames," *Environment* (1978): 8.

3. Ibid.

4. Philip Lowe and Jane Goyder, *Environmental Groups in Politics* (London: Allen & Unwin, 1983), pp. 21, 18, 20.

5. Ibid., p. 21.

6. Ibid., p. 138.

7. Ibid.

8. Stanley Johnson, *The Politics of the Environment and the British Experience* (London: Tom Stacey, 1970), p. 15.

9. Quoted in William Ashworth, *The Genesis of Modern British Town Planning* (London: Routledge & Kegan Paul, 1954), p. 230.

10. P. D. Lowe, "Amenity and Equity: A Review of Local Environmental Pressure Groups in Britain," *Environment and Planning* 9 (1977): 38.

11. Quoted in Jeremy Bugler, *Polluting Britain* (Harmondsworth: Penguin, 1972), p. 141.

12. Ann MacEwen and Malcolm MacEwen, *National Parks: Conservation or Cosmetics* (London: Allen & Unwin, 1982), p. 20.

13. Action Society Trust, *Industry and the Countryside: The Report of a Preliminary Inquiry on Amenities in the Countryside* (London: Faber & Faber, 1983), p. 182.

14. Ashby and Anderson, *Politics of Clean Air*, p. 54.

15. Howard Scarrow, "The Impact of British Domestic Air Pollution Legislation," *British Journal of Political Science* 2 (1972): 262.

16. Ashby and Anderson, *Politics of Clean Air*, p. 104.

17. Ibid., pp. 107, 109.

18. Gerald Rhodes, *Inspectorates in British Government* (London: Allen & Unwin, 1981), p. 141.

19. For a summary of the legislation, see Organization for Economic Cooperation and Development, *The State of the Environment in OECD Member Countries* (Paris, 1979), p. 125.

20. Quoted in Johnson, *Politics of the Environment*, p. 83.

21. Ibid., p. 82.

22. P. D. Lowe, "Science and Government: The Case of Pollution," *Public Administration* 54 (Autumn 1975): 287, 288.

23. P. D. Lowe, "Environmental Groups and Government in Britain," unpublished paper, p. 21. See also Lowe and Goyder, *Environmental Groups in Politics*, p. 89.

24. Francis Sandbach, "A Further Look at the Environment as a Political Issue," *International Journal of Environmental Studies* 12 (1978): 102.

25. S. N. Brookes, A. G. Jordan, R. H. Kimber, and J. J. Richardson, "The Growth of the Environment as a Political Issue," *British Journal of Political Science* (April 1976): 249, 252.

26. Quoted in Lowe and Goyder, *Environmental Groups in Politics*, p. 75.

27. Roy Gregory, "Conservation, Planning, and Politics: Some Aspects of the Contemporary British Scene," *International Journal of Environmental Studies* 4 (1972): 37.

28. "An Early Day Motion is a motion put down on the order paper by a member, theoretically for debate on an unspecified date . . . but actually to test the response of fellow members to a particular proposition. They are rarely debated, but members frequently express their support for a particular motion by subscribing their names to it. While the tabling and support of these motions is a paper exercise, . . . the volume of protest and the type of members joining may well affect a policy" (Kenneth Bradshaw and David Pring, *Parliament and Congress* [London: Quartet, 1973], p. 392).

29. Jeremy Richardson, "The Environmental Issue and the Public," in *Decision-Making in Britain*, Block V (course material, Open University, 1977), pp. 24–25.

30. Gregory, "Conservation, Planning, and Politics," p. 37.

31. William Solesbury, "The Environmental Agenda," Public Administration 54 (Winter 1976): 379.

32. Quoted in Johnson, *Politics of the Environment*, pp. 126, 127.

33. Bugler, *Polluting Britain*, p. 13.

34. Ibid., p. 48.

35. J. J. Richardson and A. G. Jordan, *Governing under Pressure* (Oxford: Martin Robertson, 1979), p. 47.

36. Bugler, *Polluting Britain*, p. 48.

37. Graham Bennett, "Pollution Control in England and Wales," *Environmental Policy and Law* 5 (1979): 95.

38. Lowe and Goyder, *Environmental Groups in Politics*, p. 144.

39. Ibid., p. 127.

40. Brian Johnson, "Ten Years Older, but Still Friends," *New Statesman*, May 3, 1981, p. 7. See also Donald Greenberg, "The Manipulation of the Media to Achieve Legitimacy: A Case Study of Britain's Friends of the Earth," paper

presented at the annual meeting of the American Political Science Association, Denver, September 2, 1980.

41. In the 1978 general election the party ran candidates in 50 parliamentary races and received a total of 40,000 votes (Chad Neighbor, "Politics with a Green Face," *Environmental Action*, November 1979, pp. 26–27.

42. Lowe and Goyder, *Environmental Groups in Politics*, pp. 42–45.

43. Lowe, "Environmental Groups and Government in Britain," p. 14.

44. Richardson and Jordan, *Governing under Pressure*, p. 68.

45. Lowe and Goyder, *Environmental Groups in Politics*, p. 66.

46. Quoted in Wyn Grant, "Insider Groups, Outsider Groups, and Interest Group Strategies in Britain," Working Paper no. 19, University of Warwick, Department of Politics, May 1970, p. 6.

47. Richardson and Jordan, *Governing under Pressure*, p. 86.

48. Grant, "Insider Groups, Outsider Groups," p. 4.

49. See Lowe and Goyder, *Environmental Groups in Politics*, pp. 62–65.

50. See Elliot J. Feldman and Jerome Milch, *Technology versus Democracy* (Boston: Auburn House, 1982).

51. Richard Kimber and J. J. Richardson, "The Roskillers: Cublington Fights the Airport," in *Campaigning for the Environment*, ed. Kimber and Richardson (London: Routledge & Kegan Paul, 1974), p. 166.

52. "The relative urgency and importance of each item [of business] before Parliament is indicated by the number of times the request for attendance is underlined by the party's floor leader—i.e., a one, two, or three-line whip—the more lines, the more the urgency. A three-line whip indicates that an important vote is expected and that a member is expected to be present" (Bradshaw and Pring, *Parliament and Congress*, p. 32).

53. Kimber and Richardson, "Roskillers," p. 168.

54. Colin Buchanan, *The Stansted Controversy—No Way to the Airport* (Essex: Longmans, 1981), p. 45.

55. Ibid., p. 46.

56. Kimber and Richardson, "Roskillers," p. 208.

57. Elliot Feldman, "British Planning and Political Instability," unpublished paper, p. 15.

58. Kimber and Richardson, "Roskillers," p. 181.

59. Richard Kimber, J. J. Richardson, and S. K. Brookes, "The Deposit of Poisonous Waste Act of 1972: A Case of Government by Reaction," *Public Law*, Autumn 1974, p. 210.

60. Ibid., p. 204.

61. Ibid.

62. Ibid., p. 209.

63. A. H. Perry, *Environmental Hazards in the British Isles* (London: Allen & Unwin, 1981), p. 139.

64. See, for example, "Tighter Controls Sought in Hazardous Waste Disposal," *Times* (London), September 9, 1981; p. 4. Also "Down in the Dumps," in Nicholas Hildyard, *Cover Up* (London: New English Library, 1981), pp. 24–50.

65. Hildyard, *Cover Up*, pp. 39–41.

66. Richard Kimber, J. J. Richardson, and S. K. Brookes, "The Juggernauts: Public Opposition to Heavy Lorries," in *Campaigning for the Environment*, ed. Kimber and Richardson, p. 139.

67. Ibid., pp. 137, 141.
68. Ibid., p. 143.
69. Ibid., pp. 144, 146.
70. Ibid., p. 151.
71. Ibid., p. 192.
72. Ibid., p. 194.
73. Ibid., p. 197.
74. Quoted in Des Wilson, *The Lead Scandal* (London: Heinemann, 1983), p. 93.
75. Ibid.
76. "The Lead Debate is Hotting Up," *Economist*, March 6, 1982, p. 97.

CHAPTER 2. *The Politics and Administration of Pollution Control*

1. In December 1982 the name of this agency was changed from Her Majesty's Alkali and Clean Air Inspectorate to Her Majesty's Industrial Air Pollution Inspectorate. Since most of the references cited in this study predate 1982, to avoid confusion I will refer to this body by its former designation and the one by which it is still commonly known, the Alkali Inspectorate.
2. Department of the Environment, Central Unit on Environmental Pollution, *Pollution Control in Britain: How It Works* (London: Her Majesty's Stationery Office, 1976), p. 6.
3. Michael Hill, "Air Pollution Control in Britain: A Case Study in Central–Local Government Reactions," unpublished paper, pp. 4–5.
4. Gerald Rhodes, *Inspectorates in British Government* (London: Allen & Unwin, 1981), p. 134.
5. Ibid., pp. 135, 136.
6. During the 1960s bitter conflicts erupted over proposals to construct new reservoirs. See, for example, Roy Gregory, "The Cow Green Reservoir," in *The Politics of Physical Resources*, ed. Peter J. Smith (Harmondsworth: Penguin, 1975), pp. 144–202.
7. See J. J. Richardson, A. G. Jordan, and R. H. Kimber, "Lobbying, Administrative Reform, and Policy Styles: The Case of Land Drainage," *Political Studies* 26 (1978): 47–64, for a discussion of the political bargaining underlying the Water Act of 1935. See also A. G. Jordan, J. J. Richardson, and R. H. Kimber, "The Origins of the Water Act of 1973," *Public Administration* 55 (1977): 318.
8. Jeremy Richardson, "The Environmental Issue and the Public," in *Decision-Making in Britain*, Block V (course material, Open University, 1977), pp. 24–25.
9. Quoted in Timothy O'Riordan, "The Role of Environmental Quality Objectives: The Politics of Pollution Control," in *Progress in Resource Management and Environmental Planning*, vol. 1, ed. Timothy O'Riordan and Ralph C. D'Arge (New York: Wiley, 1979), p. 239.
10. Department of the Environment, *Pollution Control*, p. 3.
11. Quoted in Rhodes, *Inspectorates in British Government*, p. 151.
12. Quoted in Francis Sandbach, *Principles of Pollution Control* (London: Longman, 1982), p. 74.
13. Commission of the European Communities, "Britain in the Community, 1973–1983: The Impact of Membership" (London, n.d.), p. 44.

14. Eric Ashby, "The Politics of Pollution: Reflections on the Evolution of Environmental Policy in Britain, 1820–1980," paper presented at Cornell University, 1981, p. 11.

15. Michael Hill, "The Role of the British Alkali and Clean Air Inspectorate in Air Pollution Control," in *International Comparisons in Implementing Pollution Control*, ed. Paul Downing and Kenneth Hanf (Boston: Kluwer-Nijhoff, 1983), p. 104.

16. Quoted in Sandbach, *Principles of Pollution Control*, p. 75, and in O'Riordan, "Role of Environmental Quality Objectives," p. 247.

17. Quoted in Keith Hawkins, *Environment and Enforcement: Regulation and the Social Definition of Pollution* (Oxford: Clarendon, 1984), p. 25.

18. David Storey, "An Economic Appraisal of the Legal and Administrative Aspects of Water Pollution Control in England and Wales, 1970–1974," in *Progress in Resource Management*, vol. 1, ed. O'Riordan and D'Arge, p. 268.

19. O'Riordan, "Role of Environmental Quality Objectives," p. 225.

20. Quoted in Royal Commission on Environmental Pollution, *Fifth Report* (London, 1976), p. 23.

21. Quoted in *Social Audit: A Special Report on London* (London: Social Audit Ltd., 1974), p. 8.

22. Department of the Environment, *Pollution Control in Britain*, p. 3.

23. Quoted in Sandbach, *Principles of Pollution Control*, p. 97.

24. Keith Hawkins, "The Use of Discretion by Regulatory Officials: A Case Study of Environmental Pollution in the United Kingdom," paper presented at the conference on "Perspectives on Regulation: Law, Discretion, and Bureaucratic Behavior," New York, 1982, p. 12.

25. Quoted in Christopher Wood, "Air Pollution Control," unpublished paper, p. 27.

26. Quoted in *Social Audit*, p. 5.

27. Quoted in O'Riordan, "Role of Environmental Quality Objectives," p. 225.

28. Hill, "Air Pollution Control in Britain," p. 15.

29. The following two paragraphs are based on Howard A. Scarrow, "The Impact of British Domestic Air Pollution Legislation," *British Journal of Political Science* 2 (1972): 261–82.

30. D. J. Storey, "The Economics of Environmental Law Enforcement, or Has the Prosecution of Polluters Led to Cleaner Rivers in England and Wales?" *Environment and Planning* 11 (1979): 912.

31. Quoted in Michael Hill, "Role of the British Alkali and Clean Air Inspectorate," pp. 90–91.

32. Quoted in Sandbach, *Principles of Pollution Control*, p. 96.

33. O'Riordan, "Role of Environmental Quality Objectives," p. 224.

34. Quoted in Michael Hill, "The Role of the British Alkali and Clean Air Inspectorate in Air Pollution Control," unpublished paper, p. 24.

35. Hawkins, "Use of Discretion," p. 12.

36. Quoted in Rhodes, *Inspectorates in British Government*, p. 149.

37. Quoted in Sandbach, *Principles of Pollution Control*, p. 100.

38. Quoted in Hill, "Role of the British Alkali and Clean Air Inspectorate," unpublished paper, p. 24.

39. Storey, "Economics of Environmental Law Enforcement," p. 263.

40. Jeremy Bugler, *Polluting Britain* (Harmondsworth: Penguin, 1972), p. 27.

41. Quoted in *Social Audit*, p. 15.

42. A. K. Barbour, "Environmental Processes—Problems and Solutions," speech presented to the 88th Environmental Health Congress, Harrogate, September 30, 1981, p. 9.

43. Quoted in Eric Ashby and Mary Anderson, *The Politics of Clean Air* (Oxford: Clarendon, 1981), p. 103.

44. *Social Audit*, p. 15.

45. Keith Hawkins, "Bargaining and Bluff: Compliance Strategy and Deterrence in the Enforcement of Regulation," *Law and Policy Quarterly* 5 (January 1983): 44, 45.

46. Quoted in Geoffrey Wall, "National Coping Styles: Policies to Combat Pollution," *International Journal of Environmental Studies* 9 (1976): 241.

47. Hawkins, "Bargaining and Bluff," p. 40.

48. *Social Audit*, p. 18.

49. Hawkins, "Bargaining and Bluff," p. 35.

50. *Social Audit*, p. 19.

51. Geneva Richardson, Anthony Ogus, and Paul Burrows, *Policies of Pollution: A Study of Regulation and Enforcement* (Oxford: Clarendon, 1982), p. 62.

52. Hawkins, "Bargaining and Bluff," p. 57.

53. Ibid., p. 51.

54. Quoted in *Social Audit*, p. 11.

55. Hawkins, "Bargaining and Bluff," pp. 128–29.

56. Ibid., p. 44.

57. Quoted in Timothy O'Riordan, *Environmentalism*, 2d ed. (London: Pion, 1981), p. 237.

58. Hawkins, "Bargaining and Bluff," pp. 31, 47.

59. Quoted in Hill, "Role of the British Alkali and Clean Air Inspectorate," in *International Comparisons*, ed. Downing and Hanf, p. 100.

60. Storey, "Economic Appraisal," p. 263.

61. Storey, "Economics of Environmental Law Enforcement," p. 913.

62. Quoted in Richardson, "Environmental Issues," p. 61.

63. This section is based primarily on Max Nicholson, "Economy, Society, Environment, 1920–1970," in *The Twentieth Century*, ed. Carlo M. Cipolla, Fontana Economic History of Europe, vol. 5, no. 2 (New York: Harper & Row, 1977), pp. 754–60.

64. Interview with Vivienne Kendall, association secretary.

65. Alan Irwin and Kenneth Green, "The Control of Chemical Carcinogens in Britain," *Policy and Politics* (1983): 449.

66. Nicholson, "Economy, Society, Environment," p. 757.

67. O'Riordan, "Role of Environmental Quality Objectives," p. 241.

68. O'Riordan, *Environmentalism*, pp. 232–33.

69. Quoted in Rhodes, *Inspectorates in British Government*, p. 145.

70. Quoted in Jon Tinker, "Britain's Environment—Nanny Knows Best," *New Scientist*, March 9, 1977, p. 533.

71. Ibid., p. 530.

72. Brendan Gillespie, Dave Eva, and Ron Jonston, "Carcinogens: Risk Assessment in United States and Great Britain: The Case of Aldrin/Dieldrin," *Social Studies of Science* 9 (1979): 280.

73. Christopher Wood, "Land-Use Planning and Pollution Control," in *Progress in Resource Management*, ed. O'Riordan and D'Arge, pp. 290, 294, 295.

74. Christopher Wood, "Local Planning Authority Controls over Pollution," unpublished paper, p. 8.

75. Wood, "Land-Use Planning," p. 287.

76. See Christopher Miller and Christopher Wood, *Planning and Pollution* (Oxford: Clarendon Press, 1983), p. 31.

77. Wood, "Land-Use Planning."

78. Ibid., p. 295.

79. Bugler, *Polluting Britain*.

80. Quoted in Ashby and Anderson, *Politics of Clean Air*, p. 138.

81. Quoted in O'Riordan, "Role of Environmental Quality Objectives," p. 242.

82. *Social Audit*, p. 48.

83. Quoted in Ashby and Anderson, *Politics of Clean Air*, p. 126.

84. Ibid., p. 125.

85. Tinker, "Britain's Environment," p. 530.

86. Ibid., pp. 533, 530.

87. *Social Audit*, p. 48.

88. G. M. Bates, "Public Participation under the Control of Pollution Act, 1974," *Journal of Planning and Environmental Law*, 1979, p. 210.

89. Dennis J. Parker and Edmund C. Penning-Rowsell, *Water Planning in Britain* (London: Allen & Unwin), pp. 197, 130–31.

90. Hill, "Air Pollution Control in Britain," pp. 7, 8.

91. Ibid., pp. 8, 19.

92. Royal Commission on Environmental Pollution, *Fifth Report* (London, 1976), pp. 33, 56.

93. Ibid., p. 48. For the government's response, see Department of the Environment, Central Directorate on Environmental Pollution, *Air Pollution Control*, Pollution Paper no. 18 (London, 1982).

94. Royal Commission on Environmental Pollution, *Tenth Report* (London, 1984), p. 29.

95. Tinker, "Britain's Environment," p. 530.

96. Wood, *Air Pollution Control*, p. 27.

97. A. H. Perry, *Environmental Hazards in the British Isles* (London: Allen & Unwin, 1981), p. 135.

98. Bugler, *Polluting Britain*, pp. 12, 14.

99. For a comprehensive overview of the impact of the EEC on British environmental policies, see Nigel Haigh, *EEC Environmental Policy and Britain* (London: Environmental Data Services, 1984).

100. O'Riordan, "Role of Environmental Quality Objectives," p. 249.

101. Quoted in *Social Audit*, p. 12.

102. Eric Felgate, "Industry and Pollution," *CBI Review*, 1972, p. 35.

103. Interview with CBI official.

104. Gregory S. Wetstone and Armin Rosencranz, *Acid Rain in Europe and North America: National Response to an International Problem* (Washington, D.C.: Environmental Law Institute, 1983), pp. 66–78.

105. Ibid., p. 71.

106. Ibid., p. 72.

107. Ibid., p. 74.

108. Ashby and Anderson, *Politics of Clean Air*, pp. 232–33.

109. Wetstone and Rosencranz, *Acid Rain*, p. 77.

CHAPTER 3. *The Politics and Administration of Land-Use Planning*

1. Robin Grove-White, "The Framework of Law: Some Observations," in *The Politics of Physical Resources*, ed. Peter J. Smith (Harmondsworth: Penguin/Open University Press, 1975), p. 5.

2. For an overview of the British planning system, see Anthony Barker and Mary Couper, "The Art of Quasi-Judicial Administration: The Planning, Appeal, and Inquiry Systems in England," in *Urban Law and Policy*, ed. J. P. W. McAuslan (Amsterdam: North Holland, 1985).

3. Grove-White, "Framework of Law," p. 8.

4. Quoted in Francis Sandbach, *Environment, Ideology, and Policy* (Oxford: Blackwell, 1980), p. 116.

5. Roy Gregory, *The Price of Amenity: Five Studies in Conservation and Government* (London: Macmillan, 1971), pp. 7–8.

6. Grove-White, "Framework of Law," p. 8.

7. See, for example, Roy Gregory's case study of the controversy surrounding the efforts of the Tees Valley and Cleveland Water Board to secure the passage of a private bill allowing it to build a reservoir at Cow Green: "The Cow Green Reservoir," in *Politics of Physical Resources*, ed. Smith, pp. 144–201.

8. Ian R. Manners, *North Sea Oil and Environmental Planning: The United Kingdom Experience* (Austin: University of Texas Press, 1982), p. 190.

9. John H. Noble, John S. Santa, and John S. Rosenberg, *Groping through the Maze* (Washington, D.C.: Conservation Foundation, 1977), p. 43.

10. Manners, *North Sea Oil*, p. 191.

11. Quoted in Anthony Barker, "The British 'Planning Inquiry': A Symbolic Administration Form," paper presented to the International Land Reclamation Conference, Thunock, England, April 1982, p. 3.

12. Quoted in Richard Macrory, "The Major Public Inquiry: Problems and Practices," paper presented to a Political Studies Association conference, Newcastle, April 12–14, 1983, p. 3.

13. Barker, "British 'Planning Inquiry,'" p. 3.

14. Christopher J. Duerksen, "England's Community Land Act: A Yankee's View," *Urban Law Annual* 12 (1976): 54.

15. Quoted in ibid.

16. Manners, *North Sea Oil*, pp. 192–93.

17. Barker, "British 'Planning Inquiry,'" p. 3.

18. Robin Grove-White, "Framework of Law," p. 8.

19. See Gregory, *Price of Amenity*, pp. 12–13.

20. Andrew Blowers, *The Limits of Power: The Politics of Local Planning Policy* (Oxford: Pergamon, 1980), pp. 56–57.

21. Ibid., pp. 59, 65.

22. Ibid., pp. 63, 64.

23. Ibid., p. 68.

24. Christopher Wood, "Local Planning Authority Controls over Pollution," unpublished paper, p. 18. A revised version of this paper has been published

in Christopher Miller and Christopher Wood, *Planning and Pollution* (Oxford: Clarendon, 1983), pp. 34–54.

25. Wood, "Local Planning Authority Controls," pp. 18, 19.

26. Ibid., p. 12.

27. Patrick McAuslan, *The Ideologies of Planning Law* (Oxford: Pergamon, 1980), p. 46.

28. P. D. Lowe, "Amenity and Equity: A Review of Local Environmental Pressure Groups in Britain," *Environment and Planning* 9 (1977): 38–39.

29. Philip Lowe and Jane Goyder, *Environmental Groups in Politics* (London: Allen & Unwin, 1983), p. 94.

30. William Solesbury, "The Environmental Agenda," *Public Administration* 54 (Winter 1976): 386.

31. Graham Searle, "Copper in Snowdonia National Park," in *Politics of Physical Resources*, ed. Smith, p. 75.

32. Ibid., p. 100.

33. Quoted in ibid., p. 105.

34. Ibid., p. 107.

35. Ibid., p. 111.

36. D. I. MacKay and G. A. MacKay, *The Political Economy of North Sea Oil* (London: Martin Robinson, 1975), p. 146.

37. Manners, *North Sea Oil*, p. 200.

38. MacKay and MacKay, *Political Economy*, pp. 82–83.

39. Pamela Baldwin and Malcolm E. Baldwin, *Onshore Planning for Offshore Oil: Lessons from Scotland* (Washington, D.C.: Conservation Foundation, 1975), p. 84.

40. Ibid., p. 86.

41. Ibid., p. 89.

42. Ibid.

43. Marion Shoard, *The Theft of the Countryside* (London: Temple Smith, 1980), pp. 137, 140; italics added.

44. Sandbach, *Environment, Ideology, and Policy*, p. 115.

45. Grove-White, "Framework of Law," pp. 13–14, 16.

46. Anthony Barker, "The 'Third London Airport' and Sizewell Pressured Water Nuclear Reactor: Public Inquiries and the Role of Parliament," paper presented to the annual conference of the Political Studies Association, Newcastle, April 12–14, 1983, p. 2.

47. Jeremy Bugler, *Polluting Britain* (Harmondsworth: Penguin, 1972), p. 16.

48. Anthony Blowers, "Much Ado About Nothing? A Case Study of Planning and Power," unpublished paper, p. 15. This case study constitutes part of a broader analysis of corporate power and environmental regulations that was subsequently published as *Something in the Air: Corporate Power and the Environment* (New York: Harper & Row, 1984). This controversy is also described in Jeremy Bugler, "Bedfordshire Brick," in *Politics of Physical Resources*, ed. Smith, pp. 113–43.

49. Blowers, "Much Ado," p. 16.

50. Ibid., p. 18.

51. Ibid., p. 19.

52. Ibid., pp. 20–22.

53. Ibid., p. 23.

54. Ibid., pp. 23, 24.

55. Sally M. MacGill, "Decision: Case Study, United Kingdom, Mossmorran–Braefoot Bay," International Institute for Applied Systems Analysis, A-2361, Laxenburg, Austria, July 1982, p. 95.

56. D. J. Snowball and Sally M. MacGill, "Gas Facilities in Scotland: The Assessment of Risk in Recent Siting Decisions," Working Paper no. 346, School of Geography, University of Leeds, December 1982, p. 9.

57. Ibid., pp. 11–12.

58. Ibid., p. 12.

59. MacGill, "Decision," p. 91.

60. Brian Wynne, "Windscale: A Case History in the Political Art of Muddling Through," in Progress in Resource Management and Environmental Planning, vol. 2, ed. Timothy O'Riordan and R. Kenny Turner (New York: Wiley, 1980), p. 168.

61. Francis Sandbach, "A Further Look at the Environment as a Political Issue," International Journal of Environmental Studies 12 (1978): 104.

62. Ann MacEwen and Malcolm MacEwen, National Parks: Conservation or Cosmetics? (London: Allen & Unwin, 1982), p. 235.

63. "Tighten Your Green Belt—or Not," Economist, August 15, 1983, p. 42.

64. See McAuslan, Ideologies of Planning Law, for a critique of the planning system from this perspective.

65. Sandbach, Environment, Ideology, and Policy, p. 117.

66. Ibid.

67. John Tyme, Motorways versus Democracy: Public Inquiries into Road Proposals and Their Political Significance (London: Macmillan, 1978), p. 105.

68. Sandbach, Environment, Ideology, and Policy, pp. 118–19.

69. Ibid., pp. 119, 83.

70. See Wynne, "Windscale."

71. Michael Asimov, "Delegated Legislation: United States and United Kingdom," Oxford Journal of Legal Studies 3 (1983): 264.

72. Barker, "British 'Planning Inquiry,' " pp. 6, 7.

73. David Pearce, "Public Inquiries," paper presented to the Regional Studies Association, Dundee, December 5, 1979, p. 16. For more on the present and potential role of Parliament in shaping land-use policies, see Anthony Barker, "The Legislature's Role in Official 'Decision Advice Procedures': A Proposal for Land Use Planning Policies in Britain," in Parliamentary Select Committees in Action: A Symposium, ed. Dalys M. Hill (Glasgow: Department of Politics, University of Glasgow, 1974), pp. 222–48.

74. Shoard, Theft of the Countryside, p. 102.

75. Ibid., p. 109.

76. Ibid., p. 34.

77. See "The Unacceptable Economic Face of Farming," Economist, January 15, 1983, p. 61.

78. R. H. Williams, "EEC Environmental Programme and Pollution Control," unpublished paper, pp. 6, 10. There is a considerable literature on this subject. See, for example, Peter Wathern, "The Role of Impact Statements in Environmental Planning in Britain," Institute Journal of Environmental Studies 9 (1976): 165–68; William Kennedy, "Environmental Impact Assessment: The European

Communities' Directive," discussion paper, International Institute for Environment and Society, Science Center, Berlin, 1982; and Ronald Bisset, "Quantification, Decision-Making, and Environmental Input Assessment in the United Kingdom," *Journal of Environmental Management* 7 (1978): 43–58.

79. Williams, "EEC Environmental Programme," p. 17.

80. Ibid., p. 16.

81. Nigel Haigh, "The EEC Directive on Environmental Assessment of Development Projects," paper presented to the Annual Conference of the Political Studies Association, Newcastle, April 12–14, 1983, pp. 11, 12.

CHAPTER 4. *A Comparison of British and American Environmental Regulation*

1. Organization for Economic Cooperation and Development [OECD], *The State of the Environment in OECD Member Countries* (Paris, 1979), p. 154.

2. Ibid., p. 5.

3. OECD, *Economic Implications of Pollution Control: A General Assessment* (Paris, 1974), p. 29.

4. Ibid.

5. Michael Thompson, "A Cultural Basis for Comparison," in *Risk Analysis and Decision Processes: The Siting of Liquefied Energy and Facilities in Four Countries*, ed. Howard Kunreuther and Joanne Linnerooth (Berlin: Springer, 1983), p. 233.

6. Howard C. Kunreuther and Eryl V. Ley, eds., *The Risk Analysis Controversy: An Institutional Perspective* (Berlin: Springer, 1982). For an illuminating analysis of the significance of the contrasts between British and American policy making in this area, see James Douglas, "How Actual Governments Cope with the Paradoxes of Social Choice," *Comparative Politics* 17 (1984): 67–84.

7. "Nirvana by Numbers," *Economist*, December 24, 1983, p. 56.

8. U.S. Department of Commerce, Bureau of the Census, *Statistical Abstract of the United States, 1982–83* (Washington, D.C.: Government Printing Office, 1982), p. 78.

9. Edith Efron, *The Apocalyptics* (New York: Simon & Schuster, 1984), p. 449.

10. Unless otherwise indicated, the material on air pollution in Britain is based on Jeremy Bugler, *Polluting Britain* (Harmondsworth: Penguin, 1972); Department of the Environment, *The United Kingdom Environment, 1979: Progress on Pollution Control*, Pollution Paper no. 16 (London: Her Majesty's Stationery Office, 1979), and *Digest of Environmental Pollution and Water Statistics: Pollution Control Costs* (London: Her Majesty's Stationery Office, 1981, 1982); Royal Commission on Environmental Pollution, *Fifth Report* (London, 1976), p. 40; OECD, *State of the Environment*.

11. R. E. Waller, "Air Pollution and Health," paper read at the Institute of British Atmosphere Annual Conference, University of Southampton, January 5–8, 1982, p. 2.

12. Howard Scarrow, "The Impact of British Domestic Air Pollution Legislation," *British Journal of Political Science* 2 (1972): 271.

13. A. H. Perry, *Environmental Hazards in the British Isles* (London: Allen &

Unwin, 1981), p. 132.

14. Waller, "Air Pollution and Health," p. 5.

15. Eric Ashby and Mary Anderson, *The Politics of Clean Air* (Oxford: Clarendon, 1981), pp. 141–42.

16. Perry, *Environmental Hazards*, p. 135.

17. Ibid.

18. Unless otherwise indicated, the statistics in this section are drawn from Lawrence White, *The Regulation of Air Pollutant Emissions from Motor Vehicles* (Washington, D.C.: American Enterprise Institute, 1980); Conservation Foundation, *State of the Environment: An Assessment at Mid-Decade* (Washington, D.C., 1984); annual reports of the Council on Environmental Quality, 1975–80; OECD, *State of the Environment*.

19. Conservation Foundation, *State of the Environment*, p. 87.

20. Robert Crandall, *Controlling Industrial Pollution* (Washington, D.C.: Brookings Institution, 1983), p. 19.

21. This section is based on *We Care about Water*, a booklet published by the Chemical Industry Association (n.p., n.d.), p. 5.

22. See Eric Ashby, "Legislation Outside the Factory: The British Philosophy of Pollution Control," in *Health and Industrial Growth*, CIBA Foundation Symposium 32 (Amsterdam: ASP, 1975), p. 7.

23. Perry, *Environmental Hazards*, p. 137.

24. Trevor Holloway, "The Restoration of the River Thames," *Environment* 20 (June 1978): 6.

25. Philip Revzim, "Thames Cleanup Celebrates a Success, but Acid Rain Seems to Be Tougher Task," *Wall Street Journal*, September 23, 1983, p. 32.

26. Conservation Foundation, "State of the Environment," p. 109.

27. Ibid., p. 105.

28. Council on Environmental Quality, *13th Annual Report* (Washington, D.C.: Government Printing Office, 1982), pp. 36, 37.

29. Conservation Foundation, *State of the Environment*, p. 57.

30. M. B. Green, *Pesticides: Boon or Bane?* (London: Paul Elek, 1976).

31. Richard L. Hudson, "A British Nuclear Plant Recycles Much Waste, Stirs a Growing Outcry," *Wall Street Journal*, April 11, 1984, p. 18.

32. *Investigation of the Possible Increased Incidence of Cancer in West Cambria* (London: Her Majesty's Stationery Office, 1984). See also "The Cancer Question," *Economist*, November 5, 1983, p. 57, and Hudson, "British Nuclear Plant."

33. Conservation Foundation, *State of the Environment*, p. 80.

34. See Gregory S. Wetstone and Armin Rosencranz, *Acid Rain in Europe and North America* (Washington, D.C.: Environmental Law Institute, 1983).

35. Christopher J. Duerksen, "England's Community Land Act: A Yankee's View," *Urban Law Annual* 12 (1976): 69.

36. See, for example, Nicholas Hildyard, *Cover Up* (London: New English Library, 1981).

37. Council on Environmental Quality, *13th Annual Report* (Washington, D.C.: Government Printing Office, 1982), p. 81.

38. Morris A. Ward, *The Clean Water Act: The Second Decade* (Washington, D.C.: E. Bruce Harrison, 1982), p. 3.

39. Quoted in Carter B. Lave and Gilbert S. Omenn, *Clearing the Air: Reforming the Clean Air Act* (Washington, D.C.: Brookings Institution, 1981), pp. 13, 18.

40. Lawrence J. White, *The Regulation of Air Pollutant Emissions from Motor*

*Vehicles* (Washington, D.C.: Enterprise Institute for Public Policy Research, 1982), p. 21.

41. Murray Weidenbaum, *Business, Government, and the Public*, 2d ed. (Englewood Cliffs, N.J.: Prentice-Hall, 1981), pp. 93, 94.

42. L. Thomas Galloway and J. Davitt McAteer, "Surface Mining Regulations in the Federal Republic of Germany, Great Britain, Australia, and the United States: A Comparative Study," *Harvard Environmental Law Review* 4 (1980): 278–89.

43. This paragraph is based on Lawrence White, "U.S. Mobile Source Emissions Regulations: The Problems of Implementation," in *Implementing Pollution Laws: International Comparisons*, vol. 1, ed. Paul Downing and Kenneth Hanf (Tallahassee: Florida State University, Policy Studies Program, 1981), p. 102.

44. Lennart J. Lundquist, *The Hare and the Tortoise: Clean Air Policies in the United States and Sweden* (Ann Arbor: University of Michigan Press, 1980), p. 147.

45. Walter Rosenbaum, *Environmental Politics and Policy* (Washington, D.C.: Congressional Quarterly, 1985), p. 110.

46. Council on Environmental Quality, *11th Annual Report* (Washington, D.C.: Government Printing Office, 1980), p. 109.

47. R. Shep Melnick, *Regulation and the Courts* (Washington, D.C.: Brookings Institution, 1983), p. 220.

48. This phrase is taken from the title of Christopher Stone, *Where the Law Ends* (New York: Harper & Row, 1977).

49. Helen Ingram, "The Political Rationality of Innovation: The Clean Air Act Amendments of 1970," in *Approaches to Controlling Air Pollution*, ed. Ann Friedlaender (Cambridge: MIT Press, 1978).

50. See "Clean Water: Apocalypse Later," *Regulation*, July–August 1983, pp. 9–12. (Publication of American Enterprise Institute.)

51. Quoted in Efron, *Apocalyptics*, p. 292.

52. Conservation Foundation, *State of the Environment*, pp. 63–65.

53. For a detailed account of the congressional battle over amendments to the Clean Air Act, see David Vogel, "A Case Study of Clean Air Legislation, 1967–1981," in *The Impact of the Modern Corporation*, ed. Betty Bock, Harvey J. Goldschmid, Ira M. Millstein, and F. M. Scherer (New York: Columbia University Press, 1984), pp. 309–86.

54. Melnick, *Regulation and the Courts*, p. 170.

55. Quoted in ibid., p. 173.

56. Christopher J. Duerksen, *Environmental Regulation of Industrial Plant Siting* (Washington, D.C.: Conservation Foundation, 1983), p. 56.

57. Melnick, *Regulation and the Courts*, p. 227.

58. F. E. Ireland, "The United Kingdom Viewpoint," paper presented at the International Association of Environmental Coordination Symposium, Brussels, November 16–17, 1978; Department of the Environment, *Digest of Environmental Pollution and Water Statistics*, Paper no. 4 (London: Her Majesty's Stationery Office, 1981).

59. Paul B. Downing and James K. Kimball, "Enforcing Pollution Laws in the United States," in *Implementing Pollution Laws*, ed. Downing and Hanf, p. 258.

60. According to one study, only 8 to 12% of the slowdown in productivity in the United States (between 1970 and 1979) is due to environmental regulations (Robert Havenson and Gregory B. Christiansen, "Environmental Regulations

and Productivity Growth," in *Environmental Regulation and the U.S. Economy*, ed. Henry Perkin, Paul Portney, and Allen V. Kneese [Baltimore: Resources for the Future, 1981], p. 74). In addition, studies by the Chase Manhattan Bank and Data Resources Inc. "both find the direct price, output, employment, and other micro-economic efforts of pollution control to be relatively small" (Paul Portney, "The Micro-economic Impacts of Federal Environmental Regulation," in ibid., p. 47). For an exhaustive summary of this literature, see Volkmar Hartje, "The State of Economic Research in Innovation and Environmental Protection," Papers from the International Institute for Environment and Society, Science Center, Berlin, 1984.

61. According to the OECD, the percentage of private-sector investment in pollution control in 1975 totaled 3.4 percent for the United States and 1.7 percent for the United Kingdom. The OECD cautions, however, that "these figures should not be compared without great care because of differences in statistical procedures" (*State of the Environment*, p. 134).

62. See Michael Unseem, *The Inner Circle* (New York: Oxford University Press, 1984).

63. See Timothy O'Riordan, "Public Interest Environmental Groups in the United States and Britain," *American Studies* 13 (1978): 409–38.

64. Vernon L. Smith, "On Divestiture and the Creation of Property Rights in Public Lands," *CATO Journal*, Winter 1980, p. 680.

65. For a comparative analysis, see Francis Sandbach, "A Further Look at the Environment as a Political Issue," *International Journal of Environmental Studies* 12 (1978): 99–113.

66. This analysis is developed in more detail in David Vogel, "The Public Interest Movement and the American Reform Tradition," *Political Science Quarterly* 95 (1980–81): 607–27.

67. Alfred Marcus, "Environmental Protection Agency," in *The Politics of Regulation*, ed. James Q. Wilson (New York: Basic Books, 1983), p. 288.

68. This is the central theme of Eugene Bardach and Robert Kagen, *Going by the Book* (Philadelphia: Temple University Press, 1982).

69. See, for example, ibid., pp. 49, 61, 87–88; Rosenbaum, *Environmental Politics and Policy*, p. 292; Murray Weidenbaum, *The Future of Business Regulation* (New York: AMACOM, 1979), p. 21.

70. Brendan Gillespie, Dave Eva, and Roy Johnston, "Carcinogenic Risk Assessment in the United States and Great Britain: The Case of Aldrin/Dieldrin," *Social Studies of Science* (London) 9 (1979): 293.

71. Quoted in Des Wilson, *The Lead Scandal* (London: Heinemann, 1983), p. 38.

72. Timothy O'Riordan, "The Role of Environmental Quality Objectives," in *Program in Resource Management and Environmental Planning*, ed. O'Riordan and Ralph C. D'Arge (Chichester: Wiley, 1979), 1:236–37.

73. White, *Regulation of Air Pollutant Emissions*, p. 97; Kenneth W. Chilton and Ronald J. Penoyer, *Making the Clean Air Act More Cost-Effective*, Policy Study no. 40 (St. Louis: Center for the Study of American Business, September 1981), pp. 6, 7.

74. Gillespie et al., "Carcinogenic Risk Assessment," p. 285.

75. William Renfro Havender, "Once Again Conclusive Evidence the Sky Is Falling," *Wall Street Journal*, November 3, 1983, p. 24. See also William R. Hav-

ender, "The Absence of Science in Public Policy," *Journal of Conservation Studies,* Summer 1981, pp. 5–20.

76. "Killing What?" *Economist,* June 21, 1980, p. 71.

77. "British Industry and the Environment," paper prepared by Confederation of British Industry for "Understanding British Industry, 1981."

78. Stanley Johnson, *The Politics of the Environment: The British Experience* (London: Tom Stacey, 1971), pp. 170–71.

79. O'Riordan, "Public Interest Environmental Groups," p. 252.

80. See Paul Blustein, "Dow Chemical Fights Effect of Public Outcry over Dioxin Pollution," *Wall Street Journal,* June 28, 1983, pp. 1, 24; Jeremy Main, "Dow vs. the Dioxin Monster," *Fortune,* November 30, 1983.

81. Bardach and Kagen, *Going by the Book,* p. 87.

82. Ibid.; italics in original.

83. See Robert Crandall, "Clean Air and Regional Protectionism," *Brookings Review,* Fall 1983, pp. 17–22.

84. See, for example, LeRoy Graymer and Frederick Thompson, eds., *Reforming Social Regulation* (Beverly Hills, Calif.: Sage, 1982); Lawrence White, *Reforming Regulation* (Englewood Cliffs, N.J.: Prentice-Hall, 1981); Robert Litan and William Nordhaus, *Reforming Federal Regulation* (New Haven: Yale University Press, 1983); American Enterprise Institute, *The Clean Air Act: Proposals for Revisions* (Washington, D.C., 1981); American Bar Association, *Federal Regulation: Roads to Reform,* A Report on Law and the Economy (Washington, D.C., 1978).

85. See, for example, Frederick R. Anderson, Allen V. Kneese, Phillip D. Reed, Serge Taylor, and Russell B. Stevenson, *Environmental Improvement through Economic Incentives* (Baltimore: Johns Hopkins University Press, 1977), and see Blair Bower, Rémi Barré, Jochen Kühner, and Clifford Russell, *Incentives in Water Quality Management: France and the Ruhr Area* (Washington, D.C.: Resources for the Future, 1981).

86. Christopher Wood, "Air Pollution," unpublished paper, p. 5.

87. Ibid., p. 8.

88. Chilton and Pennoyer, *Making the Clean Air Act More Effective,* pp. 15, 16.

89. Quoted in Lave and Omenn, *Clearing the Air,* pp. 26, 38–39.

90. See Duerksen, *Environmental Regulation.*

91. For more on the contrast between the "compliance" and "deterrence" forms of law enforcement, see Albert J. Reiss, Jr., "Selecting Strategies of Social Control over Organizational Life," in *Enforcing Regulation,* ed. Eugene Bardach and Robert Kagen, pp. 23–36 (Boston: Kluwer-Nijhoff, 1984).

92. Kenneth Hanf, "Some European Experiences: Suggestions for the Next Generation of Studies of Regulatory Policy," in *International Comparisons in Implementing Pollution Laws,* ed. Paul Downing and Kenneth Hanf (Boston: Kluwer-Nijhoff, 1983), p. 114.

93. Peter Knoepfel and Helmut Weidner, "Implementing Air Quality Programs in Europe: Some Results of a Comparative Study," in *International Comparisons,* ed. Downing and Hanf, p. 207.

94. This point is exhaustively documented in Melnick, *Regulation and the Courts.*

95. John Mendeloff, "Does Overregulation Cause Underregulation?" *Regulation,* September/October 1981, p. 49.

96. This is the central argument made by Keith Hawkins in *Environment and Enforcement* (Oxford: Clarendon, 1984).

CHAPTER 5. *Government Regulation in Great Britain and the United States*

1. Gary Freeman, "Do Policy Issues Determine Politics? State Pensions Policy in Britain and America," paper presented to the 1984 annual meeting of the American Political Science Association, Washington, D.C., pp. 2–3.

2. Quoted in Gerald Rhodes, *Inspectorates in British Government: Law Enforcement and Standards of Efficiency* (London: Allen & Unwin, 1981), p. 71.

3. Ibid., p. 75.

4. Quoted in ibid., p. 75.

5. See Charles D. Drake and Frank B. Wright, *Health and Safety at Work: The New Approach* (London: Sweet & Maxwell, 1983), p. 15.

6. Sheila Jasanoff, "Deregulation in the Workplace: A Comparative Analysis," unpublished paper, p. 3.

7. Rhodes, *Inspectorates in British Government*, p. 82.

8. Ibid., p. 89.

9. Jasanoff, "Deregulation in the Workplace," p. 9.

10. Quoted in Drake and Wright, *Health and Safety*, p. 14.

11. Ibid., p. 14.

12. Steven Kelman, "Enforcement of Occupational Safety and Health Regulations: A Comparison of Swedish and American Practices," in *Enforcing Regulation*, ed. Keith Hawkins and John M. Thomas (Boston: Kluwer-Nijhoff, 1984), p. 99.

13. Sheila Jasanoff, "Negotiation or Cost-Benefit Analysis: A Middle Road for U.S. Policy," *Environmental Forum*, July 1983, p. 38.

14. American Textile Manufacturers' Institute v. Donovan, 101 S.C. 2478 (1981).

15. Quoted in Joseph L. Badaracco, "Networks and Hierarchies: A Study of Business–Government Cooperation," Ph.D. dissertation, Harvard Business School, 1982, chap. 3, pp. 5, 8.

16. Ibid., p. 13.

17. Ibid., p. 24.

18. Ibid., p. 26.

19. Ibid.

20. Nicholas Hildyard, *Cover Up* (London: New English Library, 1981), pp. 169, 170.

21. "Killing What?" *Economist*, June 21, 1980, p. 71. For more on this controversy, see Hildyard, *Cover Up*, pp. 98–125, and Alan Irwin and Kenneth Green, "The Control of Chemical Carcinogens in Britain," *Policy and Politics* 11 (1983): 439–59.

22. W. G. Carson, *The Other Price of Britain's Oil: Safety and Control in the North Sea* (Oxford: Martin Robertson, 1982).

23. Graham Wilson, "Legislation on Occupational Safety and Health: A Comparison of the British and American Experience," unpublished paper, pp. 13, 12.

24. Graham Wilson, "Social Regulation and Explanations of Regulatory Failure," *Political Studies* 32 (1984): 220.

25. Ronald Brickman, Sheila Jasanoff, and Thomas Ilgen, *Chemical Regulation*

*and Cancer: A Cross-national Study of Policy and Politics* (Ithaca: Cornell University, Program on Science, Technology, and Society, 1982), p. 36.

26. Ibid., p. 39.

27. Ibid., pp. 259, 261.

28. Irwin and Green, "Control of Chemical Carcinogens," p. 443.

29. Ibid., pp. 446, 448.

30. Brickman et al., *Chemical Regulation and Cancer*, pp. 259, 380.

31. M. F. Cuthbert, J. P. Griffin, and W. H. W. Inman, "The United Kingdom," in *Controlling the Use of Therapeutic Drugs: An International Comparison*, ed. William M. Wardell (Washington, D.C.: American Enterprise Institute, 1978), p. 99.

32. Ibid., p. 103.

33. Keith Hartley and Alan Maynard, "The Regulation of the U.K. Pharmaceutical Industry: A Cost-Benefit Analysis," unpublished paper, 1981, p. 27.

34. The data in the following two paragraphs are drawn from "Testing Time for Drugs," *Economist*, August 7, 1980, pp. 69–70.

35. Quoted in David Leo Weimer, "Organizational Incentives: Safe and Available Drugs," in *Reforming Social Regulation*, ed. LeRoy Graymer and Frederick Thompson (Beverly Hills, Calif.: Sage, 1982), p. 44.

36. This section is based on Ross Cranston, *Regulating Business Law and Consumer Agencies* (London: Macmillan, 1979). For more on British consumer regulations see Peter Smith and Dennis Swann, *Protecting the Consumer* (Oxford: Martin Robertson, 1979).

37. Cranston, *Regulating Business Law*, pp. 27–28.

38. Ibid., p. 28.

39. Ibid., p. 33.

40. See "Self-Control in the City," *Economist*, July 2, 1983, p. 8.

41. This section is based on Les Metcalfe, "From Crisis Management to Preventive Medicine: The Evolution of U.K. Bank Supervision" (Berlin: International Institute of Management, 1979); "Self-Control in the City," p. 18; "Who Polices the City?" *Economist*, July 2, 1982, p. 69; "Policing the City of London," *Economist*, July 17, 1984, pp. 70–71; "A Little Cleaner," *Economist*, October 20, 1984, pp. 86–87; Barnaby Feder, "Overseeing Insurance Reform at London's Venerable Mart," *New York Times*, January 8, 1984, sec. III, p. 6.

42. R. I. Tricker, *Corporate Governance* (Aldershot: Gower, 1984), p. 23.

43. Michael Moran, *The Politics of Banking* (New York: St. Martin's, 1984), p. 117.

44. Feder, "Overseeing Insurance Reform."

45. For an extended discussion of this issue, see John Elkington, *The Ecology of Tomorrow's World* (London: Associated British Press, 1983).

46. Lennart Lundquist, *The Hare and the Tortoise: Clean Air Policies in the United States and Sweden* (Ann Arbor: University of Michigan Press, 1980).

47. Peter Knoepfel and Helmut Weidner, "Implementing Air Quality Control Programs in Europe: Some Results of a Comparative Study," in *International Comparisons in Implementing Pollution Laws*, ed. Paul Downing and Kenneth Hanf (Boston: Kluwer-Nijhoff, 1983), p. 207.

48. Brickman et al., *Chemical Regulation and Cancer*; Badaracco, "Networks and Hierarchies."

49. Steven Kelman, *Regulating America, Regulating Sweden: A Comparative Study of Occupational Safety and Health Policy* (Cambridge: MIT Press, 1981).

CHAPTER 6. *The Dynamics of Business–Government Relations in Great Britain and the United States*

1. The concept of Britain and the United States as early industrializers was first articulated by Alexander Gerschenkron in *Economic Backwardness in Historical Perspective* (Cambridge: Belknap Press of Harvard University Press, 1962), pp. 5–30.

2. Quoted in Oliver MacDonagh, "Delegated Legislation and Administrative Discretion in the 1850's: A Particular Study," *Victorian Studies*, June 1958, pp. 29–30.

3. Quoted in Harold Perkin, *The Origins of Modern English Society, 1780–1880* (London: Routledge & Kegan Paul, 1969), p. 186.

4. MacDonagh, "Delegated Legislation," p. 30.

5. See, for example, Richard Hofstadter, *Social Darwinism in American Thought* (Boston: Beacon Press, 1964), and Robert F. McClosky, *American Conservatism in the Age of Enterprise, 1865–1910* (New York: Harper Torchbooks, 1951).

6. Quoted in Perkin, *Origins of Modern English Society*, p. 225. For the American version, see Irvin G. Wyllie, *The Self-Made Man's America: The Myth of Rags to Riches* (New Brunswick, N.J.: Rutgers University Press, 1954).

7. Charles Dickens, *Hard Times* (London: Penguin Books, 1969), pp. 145–46; italics added.

8. Quoted in Neil Cunningham, *Pollution, Social Interest, and the Law* (London: Martin Robertson, 1974), pp. 70–71.

9. Oliver MacDonagh, *Early Victorian Government, 1830–1870* (London: Field & Nicolson, 1977), p. 18.

10. Joseph L. Badaracco, Jr., "Networks and Hierarchies: A Study of Business–Government Cooperation," Ph.D. dissertation, Harvard Business School, 1982, chap. 5, p. 9.

11. The extent to which British public policy during the nineteenth century can accurately be described as laissez-faire remains a subject of considerable debate among historians. See, for example, Arthur J. Taylor, *Laissez-Faire and State Intervention in Nineteenth-Century British Reality* (London: Macmillan, 1972), and R. L. Crouch, "Laissez-Faire in Nineteenth-Century Britain: Myth or Reality," *The Manchester School of Economic and Social Studies*, September 1967, pp. 199–215.

12. Quoted in Richard Abrams, "Business and Government in America," in *Scribner's Encyclopedia of American Political History* (New York, 1984), 1:130.

13. This point is well documented if little remembered. See, for example, Louis Hartz, *Economic Policy and Democratic Thought: Pennsylvania, 1776–1860* (Cambridge: Harvard University Press, 1948), and Robert Lively, "The American System: A Review Article," *Business History Review* 29 (March 1955): 81–96.

14. Andrew Shonfield, *Modern Capitalism: The Changing Balance of Public and Private Power* (New York: Oxford University Press, 1965), p. 302.

15. Quoted in Stephen Skowronek, *Building a New American State: The Expansion of National Administrative Capacities, 1877–1920* (Cambridge: Cambridge University Press), p. 135.

16. Quoted in R. Dale Grinder, "The Battle for Clean Air: The Smoke Problem in Post–Civil War America," in *Pollution and Reform in American Cities,*

*1870–1930*, ed. Martin V. Melosi (Austin: University of Texas Press, 1980), p. 92. The Minnesota Supreme Court ruled in *St. Paul v. Gilfilian* that "the emission of dense smoke from smoke stacks is not necessarily a public nuisance: whether it is or not would have to depend on the locality and surroundings" (quoted in ibid.). This argument was similar to the one made in a case decided in Britain half a century earlier: "What would be a nuisance in Belgique Square would not necessarily be so in Bermondsey" (quoted in Eric Ashby and Mary Anderson, *The Politics of Clean Air* [Oxford: Clarendon, 1981], p. 51). See also Joel Franklin Brenner, "Nuisance Law and the Industrial Revolution," *Journal of Legal Studies* 3 (1974): 403–33.

17. For a discussion of some of Britain's earliest regulatory efforts, see Peter Bartrup, "The State and the Steam-Boiler in Nineteenth-Century Britain," *International Review of Social History* 25 (1980): 77–109, and "Safety at Work: The Factory Inspectorate in the Fencing Controversy," Working Paper no. 4, Center for Socio-legal Studies, Wolfson College, Oxford, March 1979; W. G. Carson, "The Conventionalization of Early Factory Crime," *International Journal of the Sociology of Law* 7 (1979): 37–62; James M. Haas, "The Royal Dockyards: The Earliest Visitations Reform, 1749–1778," *Historical Journal* 13 (1970): 191–215.

18. For a critical view of this approach to enforcement, see W. G. Carson, "The Conventionalization of Early Factory Crime," *International Journal of the Sociology of Law* 7 (1979): 37–60. For a rebuttal, see Peter W. J. Bartrup and P. T. Fenn, "The Conventionalization of Factory Crime: A Reassessment," *International Journal of the Sociology of Law* 8 (1980): 175–86.

19. See Thomas K. McCraw, *Prophets of Regulation* (Cambridge: Harvard University Press, 1984), chap. 1.

20. A. R. Hale, "Self-Regulation and the Role of the Inspector of Factories," paper presented to the conference "Regulation in Britain," Trinity College, Oxford, September 1983, p. 7.

21. "In 1845 prosecution was a real policy instrument; by the 1860s it had become insignificant" (P. W. J. Bartrup and P. T. Fenn, "The Administration of Safety: The Enforcement Policy of the Early Factory Inspectorate, 1844–1864," *Public Administration* 58 [1980]: 95).

22. The phrase is from Perkin, *Origins of Modern English Society*, p. 252.

23. The first phrase comes from the title of Robert M. Crunden's study *Ministers of Reform* (New York: Basic Books, 1982). The second is from David Roberts, *Victorian Origins and the British Welfare State* (New Haven: Yale University Press, 1960), p. 327.

24. Bernice Martin, "The Development of the Factory Office up to 1878: Administrative Evolution and the Establishment of a Regulatory Style in the Early Factory Inspectorate," paper presented to the conference "Regulation in Britain," Trinity College, Oxford University, September 1983, p. 6.

25. Roberts, *Victorian Origins*, p. 230.

26. See, for example, *The Muckrakers*, ed. Arthur Weinberg and Lila Weinberg (New York: Capricorn, 1964).

27. Roberts, *Victorian Origins*, p. 152.

28. Quoted in Perkin, *Origins of Modern English Society*, pp. 323–24.

29. Herbert Croly, "A Philosophy for Reform," in *Progressivism: The Central Issues*, ed. David Kennedy (Boston: Beacon Press, 1971).

30. Quoted in Perkin, *Origins of Modern English Society*, p. 450.

31. See, for example, S. E. Finer, "The Transmission of Benthamite Ideas,

1982–1950," in *Studies in the Growth of Nineteenth-Century Government*, ed. Gillian Sutherland (Totowa, N.J.: Rowman & Littlefield, 1971); L. J. Hume, "Jeremy Bentham and the Nineteenth-Century Revolution in Government," *Historical Journal* 10 (1967): 361–75; and David Roberts, "Jeremy Bentham and the Victorian Administrative State," *Victorian Studies* 2 (March 1959): 190–210.

32. Quoted in Henry Parris, "The Nineteenth-Century Revolution in Government: A Reappraisal Reappraised," in *The Victorian Revolution*, ed. Peter Stansky (New York: New Viewpoints, 1975), p. 33.

33. Oliver MacDonagh, "The Nineteenth-Century Revolution in Government: A Reappraisal," in *The Victorian Revolution*, ed. Stansky, p. 19.

34. Ingeborg Paulus, *The Search for Pure Food* (London: Martin Robinson, 1974).

35. Harry Eckstein, "The British Political System," in *Patterns of Government*, ed. Samuel Beer and Adam B. Ulam, 2d ed. (New York: Random House, 1962), p. 207.

36. Donald Read, *England, 1868–1914* (London: Longman's, 1979), p. 171.

37. See Asa Briggs, *Victorian Cities* (Harmondsworth: Penguin, 1968).

38. Quoted in Perkin, *Origins of Modern English Society*, p. 450; italics in original.

39. Martin, "Development of the Factory Office," pp. 8, 10.

40. Badaracco, "Networks and Hierarchies," chap. 3, p. 8.

41. Quoted in Morrell Heald, *The Social Responsibilities of Business: Company and Community, 1900–1960* (Cleveland: Press of Case Western Reserve University, 1970), p. 32.

42. Walter Lippmann, *Drift and Mastery* (Englewood Cliffs, N.J.: Prentice-Hall, 1961), pp. 22, 23.

43. Samuel Hays, *Conservation and the Gospel of Efficiency* (New York: Atheneum, 1959); Gabriel Kolko, *The Triumph of Conservatism* (Chicago: Quadrangle, 1967); James Weinstein, *The Corporate Ideal in the Liberal State, 1900–1918* (Boston: Beacon Press, 1968); Robert N. Wiebe, *Businessmen and Reform* (Chicago: Quadrangle, 1962).

44. Quoted in Samuel P. Hays, "Who Were the Progressives?" in *Progressivism*, ed. Kennedy, p. 91.

45. Quoted in ibid., p. 106.

46. Skowronek, *Building a New American State*, p. 91.

47. Dewey W. Grantham, Jr., "The Progressive Era and the Reform Tradition," in *Progressivism*, ed. Kennedy, p. 197.

48. Gabriel Kolko, "The Triumph of Conservatism," in *Progressivism*, ed. Kennedy, p. 136.

49. See, for example, Kolko, *Triumph of Conservatism*, pp. 98–108.

50. Quoted in Eugene Bardach and Robert Kagan, *Going by the Book* (Philadelphia: Temple University Press, 1982), p. 38.

51. For a detailed account of the conflict among particular industries over social regulation, see Donna Wood, *Strategic Uses of Public Policy: Business and Government in the Progressive Era* (Boston: Pitman, 1985).

52. Grinder, "Battle for Clean Air."

53. Ibid., p. 100.

54. This account is based on A. E. Dingle, " 'The Monster Nuisance of All': Landowners, Alkali Manufacturers, and Air Pollution, 1828–1864," *Economic History Review* 35 (November 1982): 529–48; Ashby and Anderson, *Politics of*

*Clean Air*, pp. 20–32; Francis Sandbach, "Pollution Control: In Whose Interest?" unpublished paper; and Roy MacLeod, "The Alkali Acts Administration, 1963–84: The Emergence of the Civil Scientist," *Victorian Studies* 9 (December 1965): 85–112.

55. Dingle, " 'Monster Nuisance,' " p. 545.

56. Ashby and Anderson, *Politics of Clean Air*, pp. 25–26.

57. Quoted in ibid., p. 28.

58. Quoted in Maurice Frankel, "The Alkali Inspectorate," *Social Audit*, Spring 1974, p. 5.

59. The literature on social regulation between the Progressive Era and the 1960s is thin. See, for example, Joseph Pratt, "Letting the Grandchildren Do It: Environmental Planning during the Ascent of Oil as a Major Energy Source," *Public Historian* (Summer 1980): 28–61; and Joel A. Tarr and Bill C. Lamperes, "Changing Fuel Use Behavior and Energy Transitions: The Pittsburgh Smoke Control Movement, 1940–1950," *Journal of Social History* 14 (Summer 1981): 561–88.

60. Martin Weiner, *English Culture and the Decline of the Industrial Spirit, 1850–1980* (Cambridge: Cambridge University Press, 1981), p. 26.

61. Ibid., p. 14.

62. Ibid., p. 23.

63. Quoted in Ashby and Anderson, *Politics of Clean Air*, p. 28.

64. Daniel Bell, "The Future That Never Was," *Public Interest* 51 (1978): 35–73.

65. Joseph A. Schumpeter, *Capitalism, Socialism, and Democracy*, 3d ed. (New York: Harper Torchbooks, 1942), p. 136; Schumpeter's italics.

66. Skowronek, *Building a New American State*, p. 55.

67. Robert B. Reich, *The Next American Frontier* (New York: Times Books, 1983), p. 33.

68. For a detailed discussion and analysis of this theme in American business ideology, see David Vogel, "Why Businessmen Distrust Their State: The Political Consciousness of American Corporate Executives," *British Journal of Political Science* 8 (January 1978): 45–78.

69. Louis Hartz, *The Liberal Tradition in America* (New York: Harcourt, Brace & World, 1951), p. 232.

70. This point is documented in Thomas McCraw, "Business and Government: The Origins of the Adversary Relationship," *California Management Review*, Winter 1984, pp. 33–52.

71. Samuel P. Huntington, *American Politics and the Promise of Disharmony* (Cambridge: Harvard University Press, 1981).

72. Weiner, *English Culture*, pp. 138–39.

73. See James Prothro, *Dollar Decade* (Baton Rouge: Louisiana State University Press, 1951); see also Ellis Hawley, "Three Facets of Hooverian Associationalism," in *Regulation in Perspective*, ed. Thomas McCraw (Cambridge: Harvard University Press, 1981), pp. 95–123.

74. See, for example, Derek Fraser, *The Evolution of the British Welfare State* (London: Macmillan, 1973), and Paul Addison, *The Road to 1945: British Politics and the Second World War* (London: Quartet, 1973).

75. See Eckstein, "British Political System," p. 209.

76. There is a large and depressing literature on this subject. See, for example, Stephen Young with A. V. Lowe, *Intervention in the Mixed Economy* (Lon-

don: Croom Helm, 1974); Trevor Smith, *The Politics of the Corporate Economy* (Oxford: Martin Robertson, 1979).

77. J. P. Nettl, "Consensus or Elite Domination: The Case of Business," *Political Studies* 12 (1965): 23.

78. Quoted in Shonfield, *Modern Capitalism.*

79. See, for example, Jonathan Boswell, "The Informal Social Control of Business in Britain: 1880–1939," *Business Historical Review* 57 (Summer 1983): 237–57. For a different analysis of the meaning of corporate social responsibility in Britain and America, see Edwin M. Epstein, "The Social Role of Business Enterprise in Britain," pt. 1, *British Journal of Management Studies*, October 1976, pp. 217–33, and pt. 2, ibid., October 1977, pp. 281–316.

80. For a more detailed discussion of this development, see David Vogel, "The Political Power of Business in America: A Reappraisal," *British Journal of Political Science* 13 (1983): 19–43.

81. Murray Weidenbaum, "The Second Managerial Revolution: The Shift of Economic Decision-making from Business to Government," in *Planning, Politics, and the Public Interest*, ed. Walter Goldstein (New York: Columbia University Press, 1978).

82. Gladwin Hill, "The Politics of Air Pollution: Public Interest and Pressure Groups," *Arizona Law Review* 10 (Summer 1968): 41.

83. Thomas W. Benham, "Trends in Public Attitudes toward Business and the Free Enterprise System," paper presented to the White House conference "The Industrial World Ahead," Washington, D.C., 1971, pp. 13–14.

84. Arthur Downs, "Up and Down with Ecology: The Issue Attention Cycle," *Public Interest*, no. 28 (Summer 1972), p. 27.

85. For a more detailed account of the ideology of the public interest movement, see David Vogel, "The Public Interest Movement and the American Reform Tradition," *Political Science Quarterly* 95 (Winter 1980–81).

86. See Lawrence White, *The Regulation of Air Pollutant Emissions from Motor Vehicles* (Washington, D.C.: American Enterprise Institute, 1982).

87. See Edith Efron, *The Apocalyptics* (New York: Simon & Schuster, 1984).

88. Mary Douglas and Aaron Wildavsky, *Risk and Culture* (Berkeley: University of California Press, 1982); italics added.

89. Ben Wattenberg, *The Good News Is the Bad News Is Wrong* (New York: Simon & Schuster, 1984), p. 32.

90. See Howard Kunreuther and Eryl V. Ley, *The Risk Analysis Controversy* (Berlin: Springer, 1982), and "Ratin and Reagan," *Environmental Action*, July–August 1983, p. 8.

91. Quoted in Keith Hawkins and John Thomas, "The Enforcement Process in Regulatory Bureaucracies," in Hawkins and Thomas, *Enforcing Regulations* (Boston: Kluwer-Nijhoff, 1984), pp. 4–5.

92. Henry Steck, "Why Does Industry Always Get What It Wants?" *Environmental Action*, July 24, 1981.

93. Christopher Wood, "Planning Pollution Prevention," unpublished manuscript, chap. 4, p. 2; quoted by permission.

94. "Forcing Technology: The Clean Air Act Experience," *Yale Law Journal* 88 (1979): 1718, 1719.

95. See R. Shep Melnick, "Pollution Deadlines and the Coalition for Failure," *Public Interest*, no. 75 (Spring 1984), pp. 123–34.

96. J. Clarence Davies III and Barbara S. Davies, *The Politics of Pollution*, 2d ed. (Indianapolis: Bobbs-Merrill, 1975), p. 53.

97. Quoted in Frank V. Fowlkes, "Washington Processes," *National Journal*, November 14, 1970, p. 2507.

98. Norman Vig and Michael Kraft, *Environmental Policy in the 1980's* (Washington, D.C.: Congressional Quarterly, 1984), p. 285.

99. Quoted in ibid., pp. 285–86.

100. Ibid., p. 285.

101. Quoted in Walter Rosenbaum, *Environmental Politics and Policy* (Washington, D.C.: Congressional Quarterly, 1985), p. 296.

102. Quoted in Joan Claybrook, *Retreat from Safety* (New York: Pantheon, 1984), p. 99.

103. See, for example, Michael Winer, "Auchter's Record at OSHA Leaves Labor Outraged, Business Satisfied," *National Journal*, October 1, 1983.


CHAPTER 7. *Government Regulation and Comparative Politics*

1. Alexander Gerschenkron, "Economic Backwardness in Historical Perspective," in *Economic Backwardness in Historical Perspective*, ed. Gerschenkron (Cambridge: Belknap Press of Harvard University Press, 1961), p. 7.

2. Andrew Shonfield, *Modern Capitalism* (New York: Oxford University Press, 1965).

3. Philippe Schmitter, "Still the Century of Corporatism?" in *Trends toward Corporatist Intermediation*, ed. Philippe Schmitter and Gerhard Lehmbruch, pp. 7–52 (Beverly Hills, Calif.: Sage, 1979).

4. Stephen Krasner, "United States Commercial and Monetary Policy," in *Between Power and Plenty*, ed. Peter Katzenstein (Madison: University of Wisconsin Press, 1978), p. 57.

5. Peter J. Katzenstein, *Small States in World Markets: Industrial Policy in Europe* (Ithaca: Cornell University Press, 1985), p. 20.

6. John Zysman, *Governments, Markets, and Growth: Financial Systems and the Politics of Industrial Change* (Ithaca: Cornell University Press, 1983), p. 7.

7. Schmitter, "Still the Century of Corporatism?" p. 24.

8. Joseph L. Badaracco, Jr., "Networks and Hierarchies: A Study of Business–Government Cooperation," Ph.D. dissertation, Harvard Business School, 1982; Ronald Brickman, Sheila Jasanoff, and Thomas Ilgen, *Chemical Regulation and Cancer* (Ithaca: Cornell University, Program on Science, Technology, and Society, 1982).

9. Zysman, *Governments, Markets, and Growth*, p. 297.

10. Samuel H. Beer, *British Politics in the Collectivist Age* (New York: Vintage, 1969), p. 329.

11. Harry Eckstein, *Pressure Group Politics* (Stanford: Stanford University Press, 1960), p. 24.

12. Grant Jordan and Jeremy Richardson, "The British Policy Style or the Logic of Negotiation?" in *Policy Styles in Western Europe*, ed. Jeremy Richardson (London and Boston: Allen & Unwin, 1982), pp. 81, 85.

13. Kenneth Waltz, *Foreign Policy and Democratic Politics* (Boston: Little, Brown, 1967), pp. 31–32.

14. Quoted in Jordan and Richardson, "British Policy Style," p. 81.

15. David Kirp, *Doing Good by Doing Little* (Berkeley: University of California Press, 1979), pp. 55–56.

16. Eckstein, *Pressure Group Politics*, p. 24.

17. Claus Offe, "The Attribution of Public Status to Interest Groups: Observations on the West German Case," in *Organizing Interests in Western Europe*, ed. Suzanne Berger (Cambridge: Cambridge University Press, 1981), p. 135.

18. J. J. Richardson and A. G. Jordan, *Governing under Pressure* (Oxford: Martin Robertson, 1974), p. 100.

19. Ibid., p. 69.

20. Ibid., pp. 116–17.

21. Grant McConnell, *Private Power and American Democracy* (New York: Vintage, 1966), p. 7.

22. Theodore Lowi, *The End of Liberalism* (New York: Norton, 1969).

23. See, for example, Marvin Bernstein, *Regulating Business by Independent Commission* (Princeton: Princeton University Press, 1955).

24. Harry Eckstein, "The Sources of Leadership and Democracy in Britain," in *Patterns of Government*, ed. Samuel Bear and Adam Ulman, 2d ed. (New York: Random House, 1966), p. 172.

25. See Thomas Gair, Mark Peterson, and Jack Walker, "Interest Groups, Iron Triangles, and Representative Institutions in American National Government," *British Journal of Political Science* 14 (1984): 161–62.

26. Robert Salisbury, "Why No Corporatism in America?" in *Trends toward Corporatist Intermediation*, ed. Schmitter and Lehmbruch, p. 221.

27. The significance of the differences in the relationships between politicians and civil servants in Great Britain and the United States is developed in more detail in James Douglas, "How Actual Governments Cope with the Paradoxes of Social Choice," *Comparative Politics* 7 (1984): 67–84.

28. Eckstein, *Pressure Group Politics*, p. 18.

29. The first quotation is from ibid.; the latter is from Salisbury, "Why No Corporatism in America?" p. 221.

30. Salisbury, "Why No Corporatism in America?" p. 220.

31. Eckstein, *Pressure Group Politics*, p. 24.

32. Michael Asimov, "Delegated Legislation: United States and United Kingdom," *Oxford Journal of Legal Studies* 3 (1983): 273–74.

33. See OECD, *Environmental Policies in Japan* (Paris, 1977).

34. These paragraphs are based on the following sources: "Nirvana by Numbers," *Economist*, December 24, 1983, p. 56; Edith Efron, *The Apocalyptics* (New York: Simon & Schuster, 1984), p. 449; and John Braithwaite, *To Punish or Persuade* (Albany: State University of New York Press, 1985), pp. 1–6.

35. David Handley, "L'Ecologie et les attitudes politiques de France," unpublished paper, University of Geneva.

36. For Germany, see Elim Papadakis, *The Green Movement in Germany* (New York: St. Martin's, 1984); for Japan, see Margaret A. McKean, *Environmental Protest and Citizen Politics in Japan* (Berkeley: University of California Press, 1981).

37. Peter Hall, "The Role of the State in the Decline of the British Economy," unpublished paper, p. 16.

38. Zysman, *Governments, Markets, and Growth*, p. 202.

39. Asimov, "Delegated Legislation," p. 269; Asimov's italics.

# Index

*Library of Congress Cataloging-in-Publication Data*

Vogel, David, 1947–
  National styles of regulation.

  Includes index.
    1. Environmental policy—Great Britain. 2. Environmental policy—United States. 3. Business and politics—Great Britain. 4. Business and politics United States. 5. Industry and state—Great Britain. 6. Industry and state—United States. I. Title.
  HC260.E5V64   1986        363.7′056′0941        85–21332
  ISBN 0–8014–1658–2
  ISBN 0–8014–9353–6 (pbk.)